I0489646

Number of the Day Activity Book

Beyond 100

Margaret Brown

What is the number of the day?

Write the number using words:

Write the number three different ways using sticks and circles.
(Remember a stick represents a ten and a circle represents a one. If you know how to use a square block to represent a hundred, you may do so.)

Fill in the correct tens and ones for the given numbers.

☐	tens and	☐	ones =	37
☐	tens and	☐	ones =	43
☐	tens and	☐	ones =	81
☐	tens and	☐	ones =	72

Find your way out of the maze.

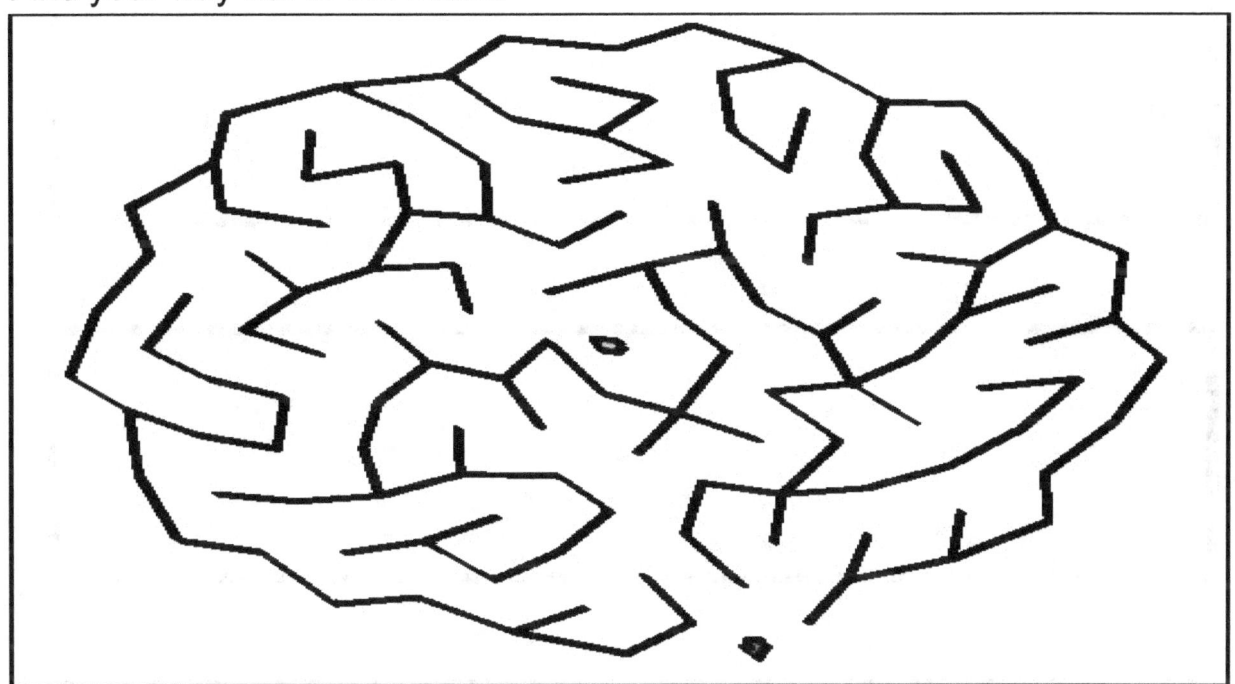

What is the number of the day?

Write the number using words:

Write the number three different ways using sticks and circles.
(Remember a stick represents a ten and a circle represents a one. If you know how to use a square block to represent a hundred, you may do so.)

Fill in the correct tens and ones for the given numbers.

	tens and		ones	=	44
	tens and		ones	=	56
	tens and		ones	=	87
	tens and		ones	=	10

Find your way out of the maze.

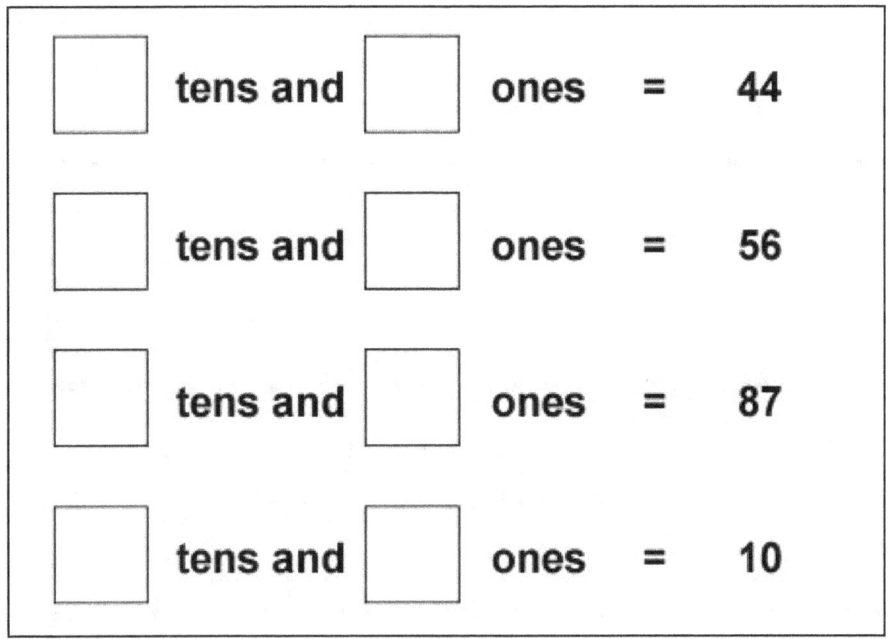

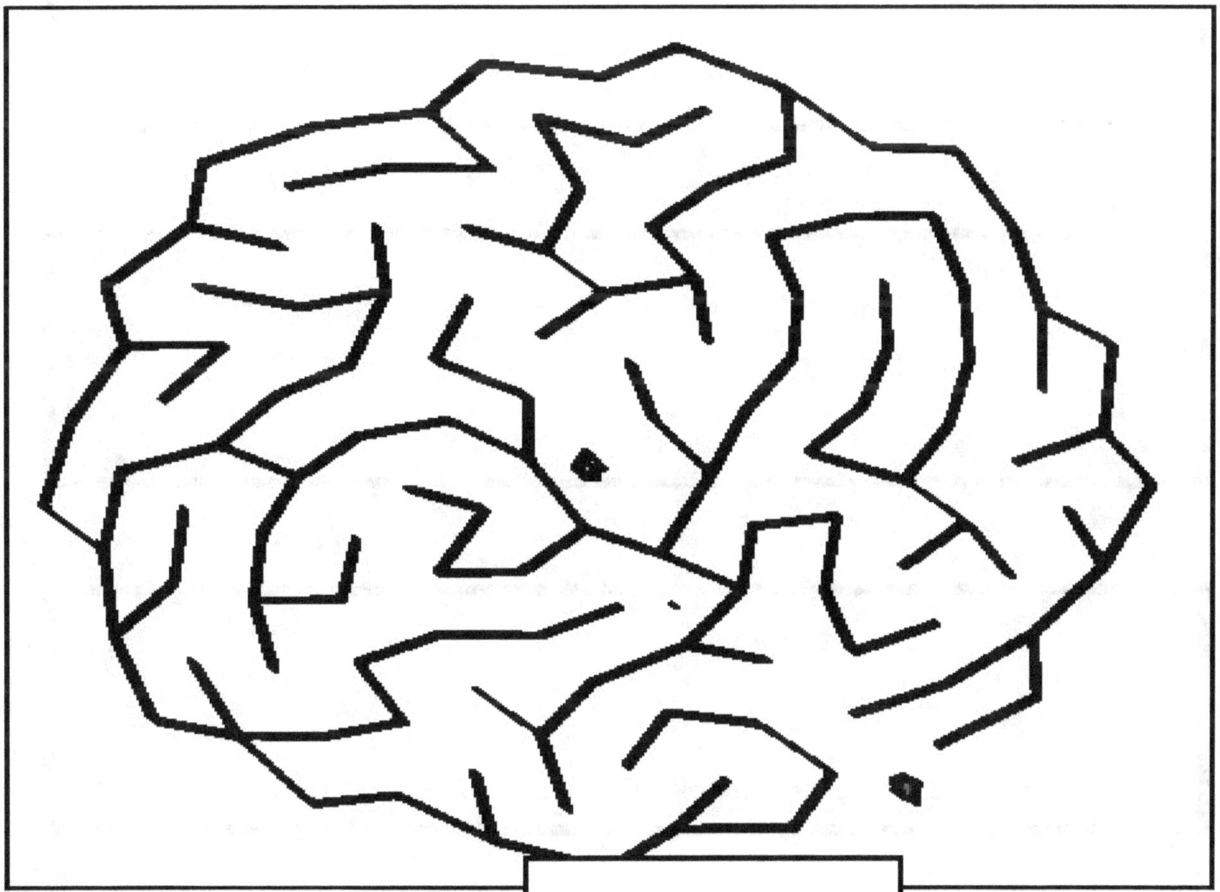

What is the number of the day?

Write the number using words:

Write the number three different ways using sticks and circles.
(Remember a stick represents a ten and a circle represents a one. If you know how to use a square block to represent a hundred, you may do so.)

Fill in the correct tens and ones for the given numbers.

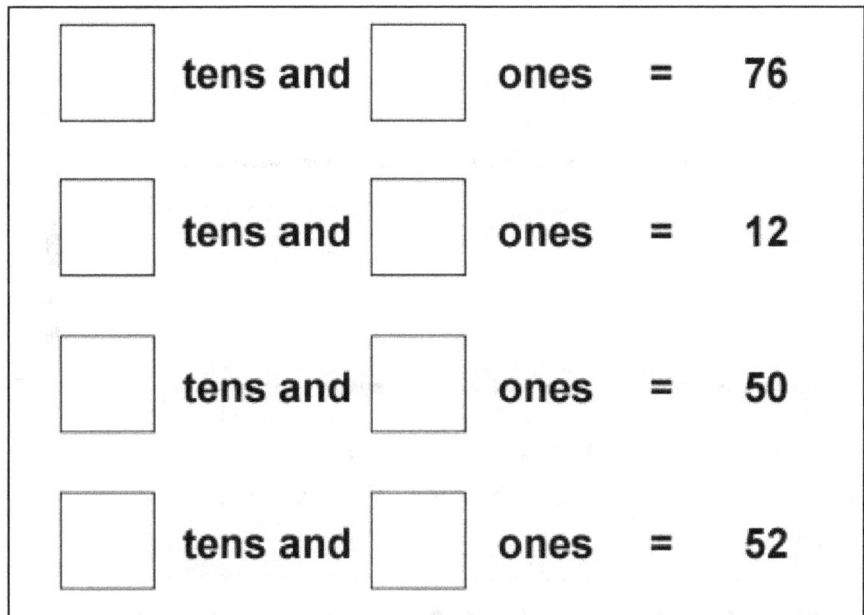

	tens and		ones	=	76
	tens and		ones	=	12
	tens and		ones	=	50
	tens and		ones	=	52

Find your way out of the maze.

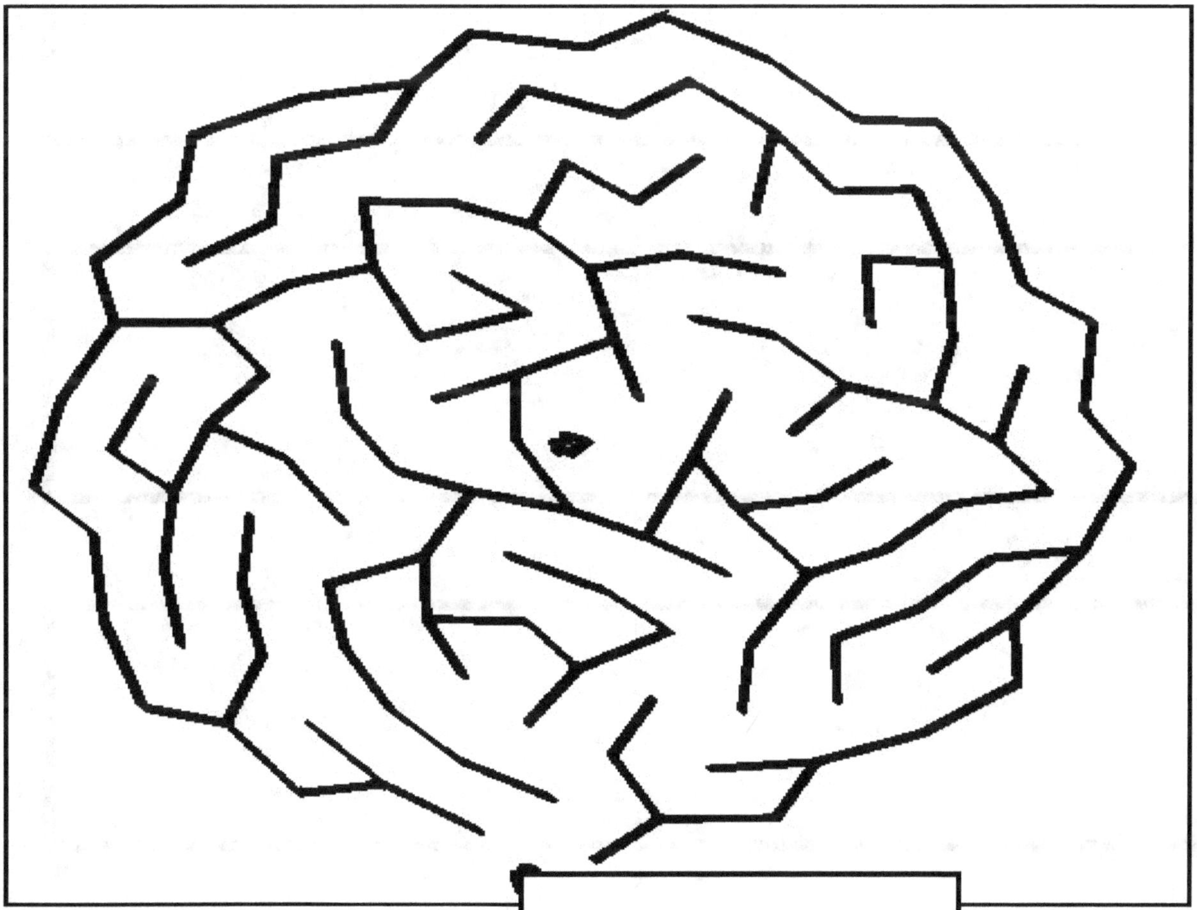

What is the number of the day?

Write the number using words:

[ ]

Write the number three different ways using sticks and circles.
(Remember a stick represents a ten and a circle represents a one. If you know how to use a square block to represent a hundred, you may do so.)

[ ]

[ ]

[ ]

Fill in the correct tens and ones for the given numbers.

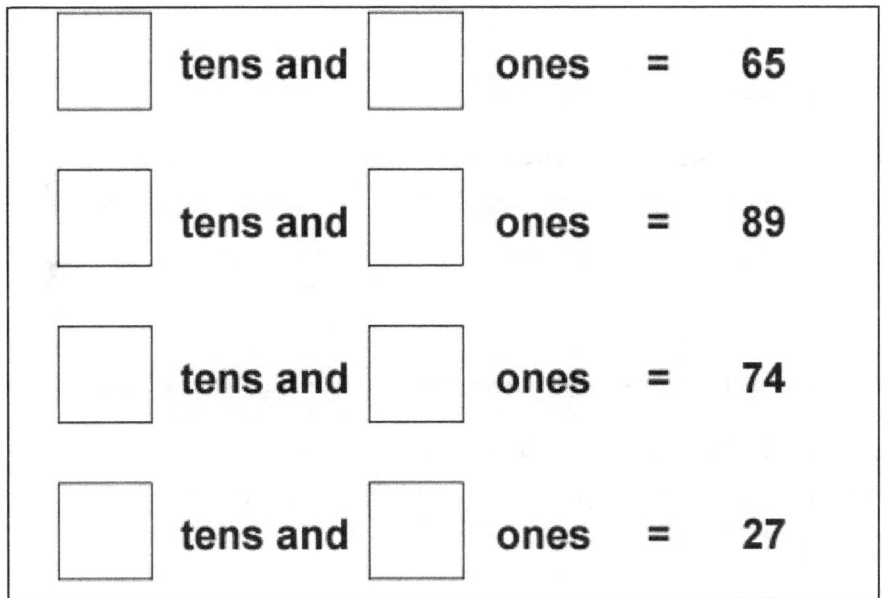

	tens and		ones	=	65
	tens and		ones	=	89
	tens and		ones	=	74
	tens and		ones	=	27

Find your way out of the maze.

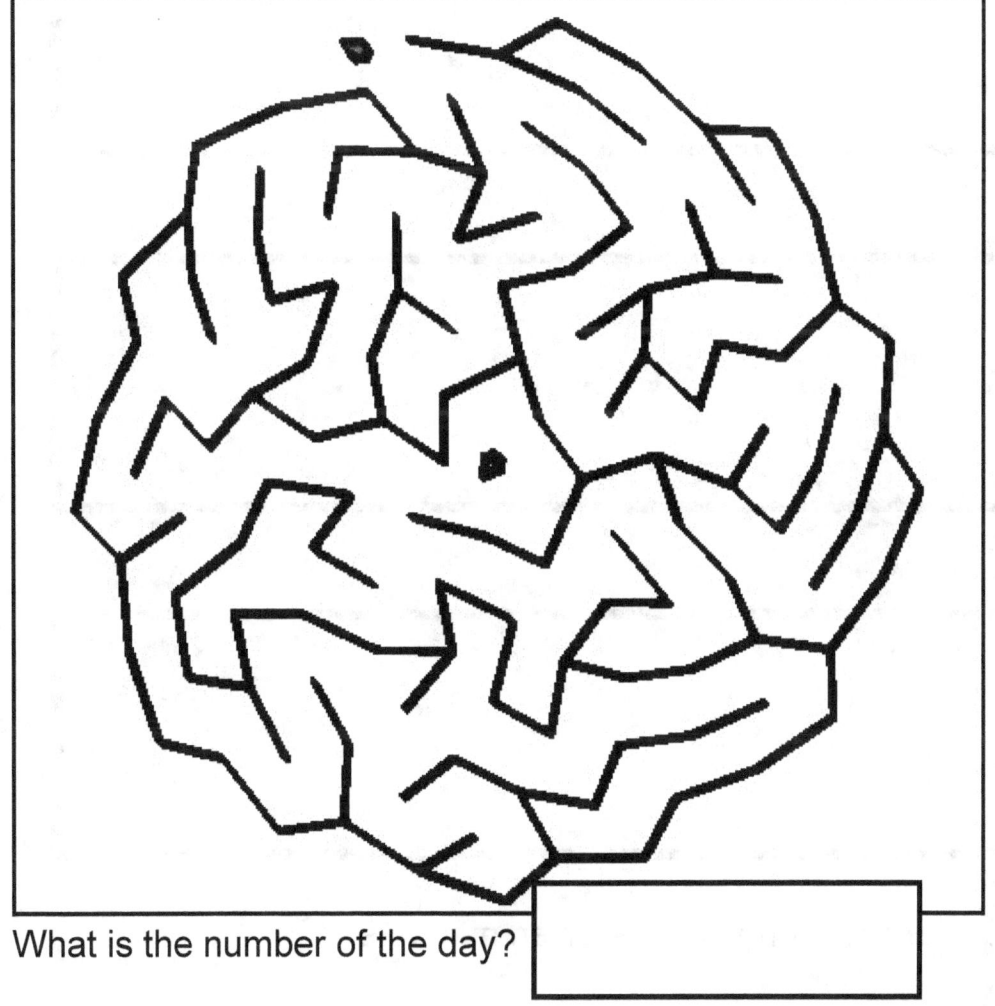

What is the number of the day?

Write the number using words:

Write the number three different ways using sticks and circles.
(Remember a stick represents a ten and a circle represents a one. If you know how to use a square block to represent a hundred, you may do so.)

Fill in the correct tens and ones for the given numbers.

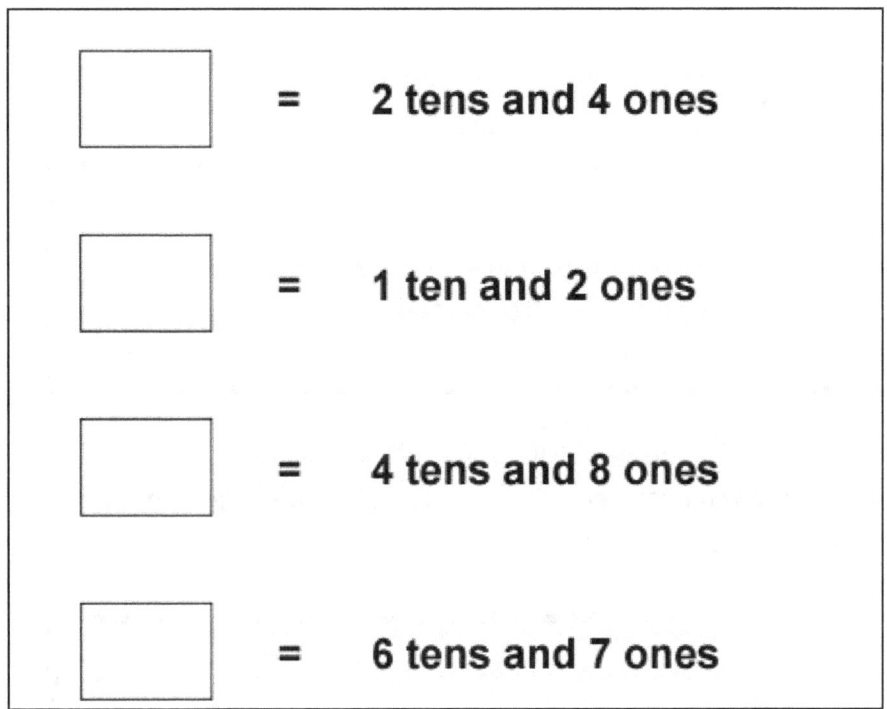

	=	2 tens and 4 ones
	=	1 ten and 2 ones
	=	4 tens and 8 ones
	=	6 tens and 7 ones

Find your way out of the maze.

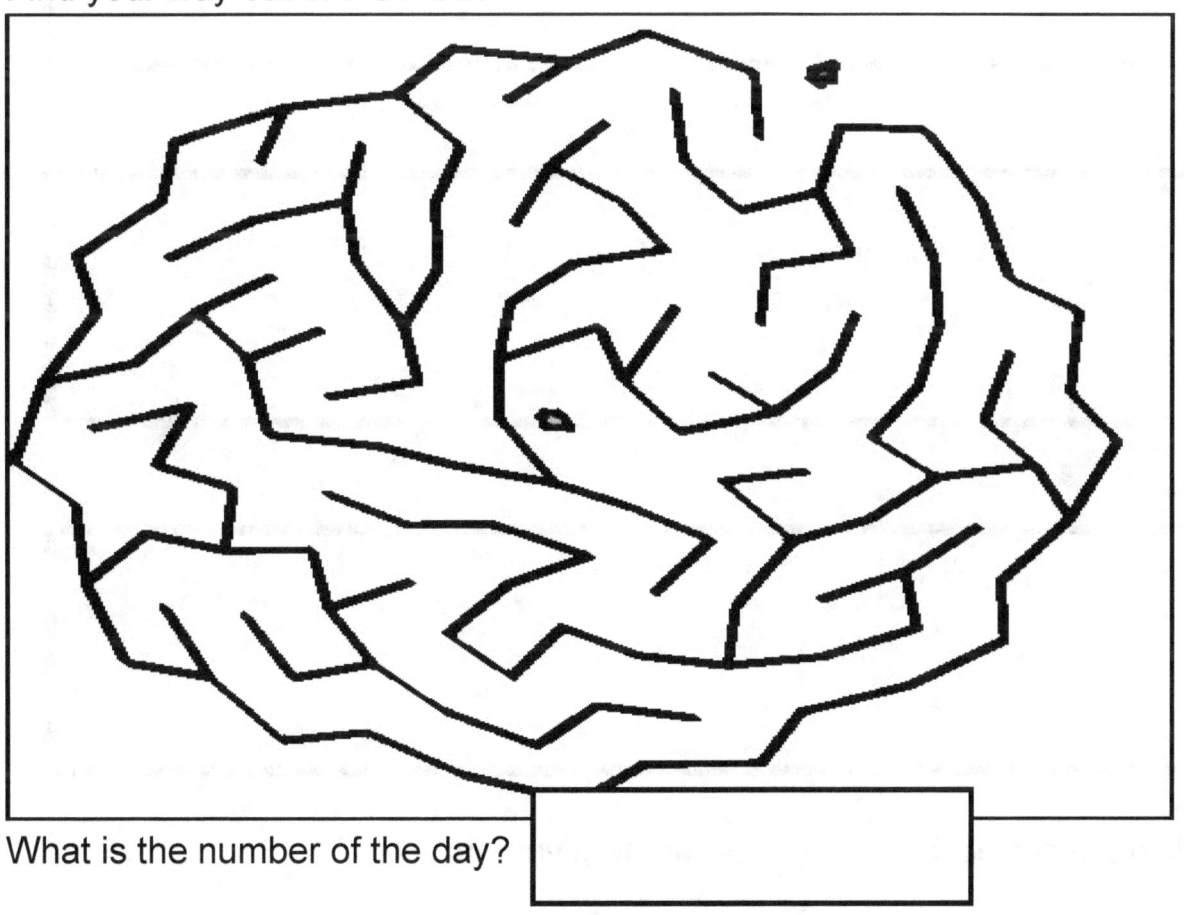

What is the number of the day?

Write the number using words:

Write the number three different ways using sticks and circles.
(Remember a stick represents a ten and a circle represents a one. If you know how to use a square block to represent a hundred, you may do so.)

Fill in the correct tens and ones for the given numbers.

	=	**5 tens and 5 ones**
	=	**9 tens and 6 ones**
	=	**8 tens and 0 ones**
	=	**3 tens and 9 ones**

Find your way out of the maze.

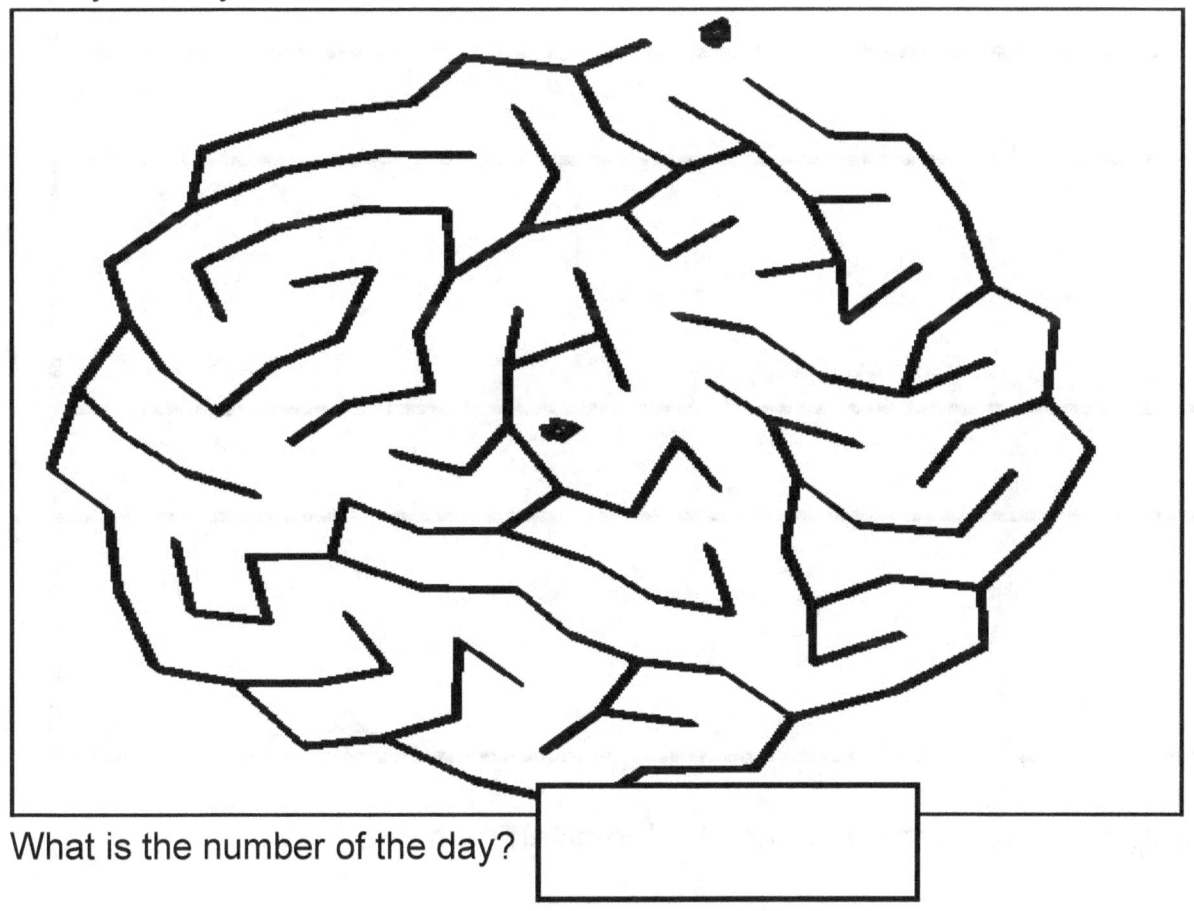

What is the number of the day?

Write the number using words:

[blank box]

Write the number three different ways using sticks and circles.
(Remember a stick represents a ten and a circle represents a one. If you know how to use a square block to represent a hundred, you may do so.)

[blank box]

[blank box]

[blank box]

Fill in the correct tens and ones for the given numbers.

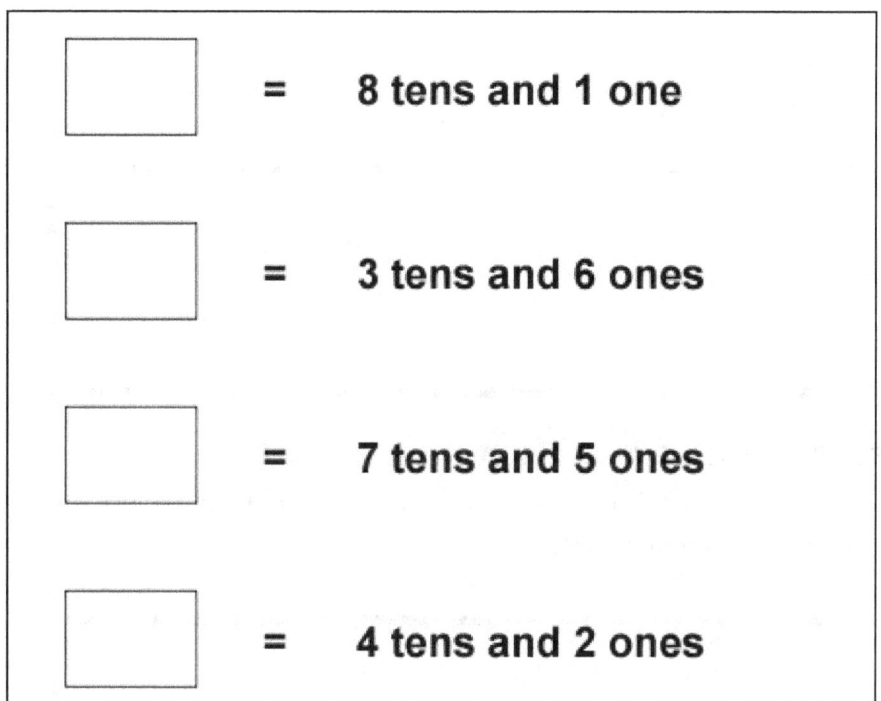

	=	8 tens and 1 one
	=	3 tens and 6 ones
	=	7 tens and 5 ones
	=	4 tens and 2 ones

Find your way out of the maze.

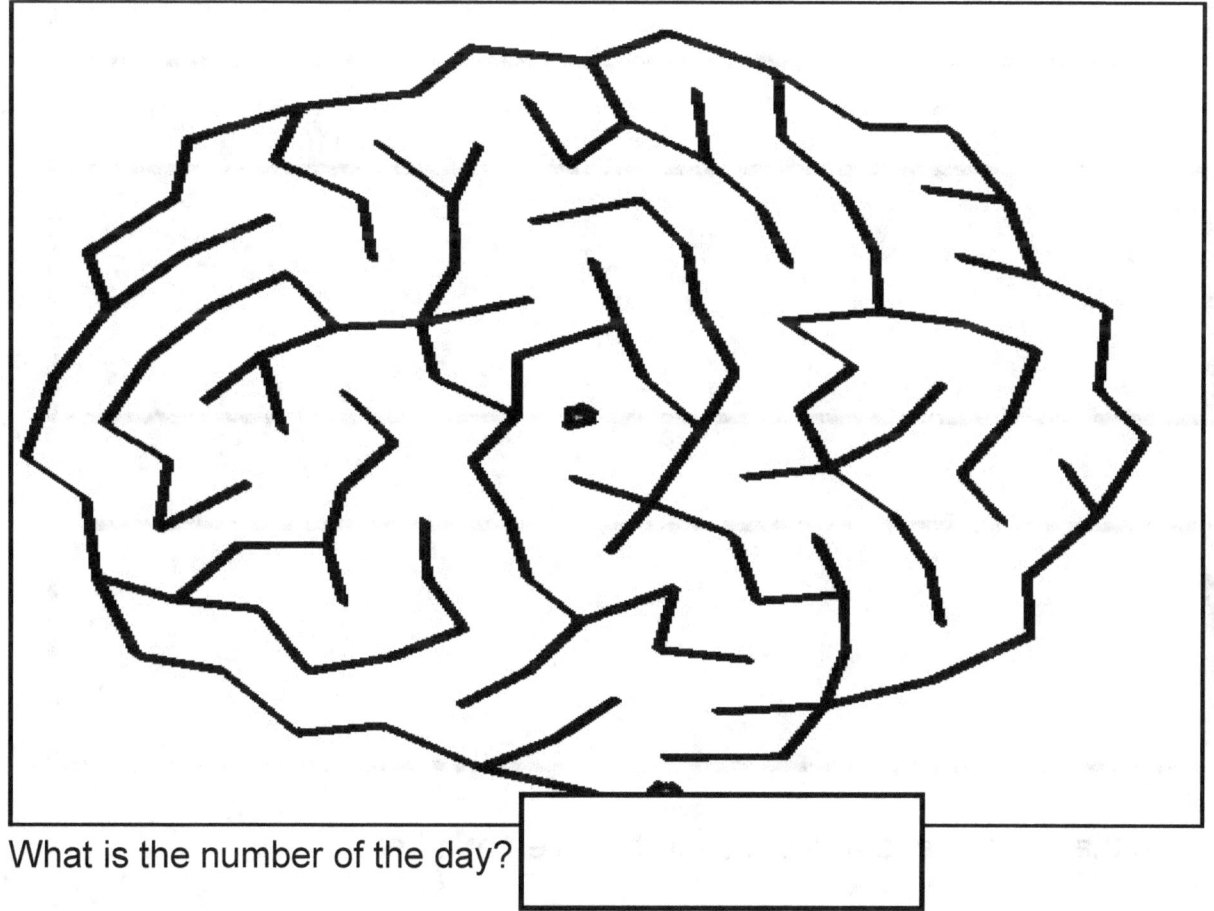

What is the number of the day?

17

Write the number using words:

Write the number three different ways using sticks and circles.
(Remember a stick represents a ten and a circle represents a one. If you know how to use a square block to represent a hundred, you may do so.)

Fill in the correct tens and ones for the given numbers.

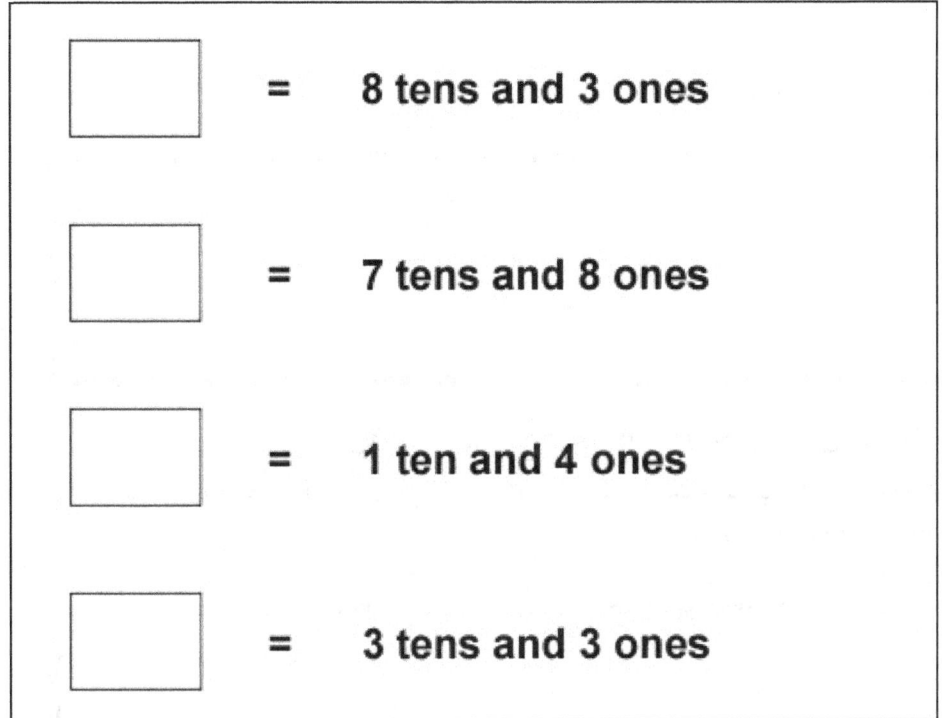

[ ] = 8 tens and 3 ones

[ ] = 7 tens and 8 ones

[ ] = 1 ten and 4 ones

[ ] = 3 tens and 3 ones

Find your way out of the maze.

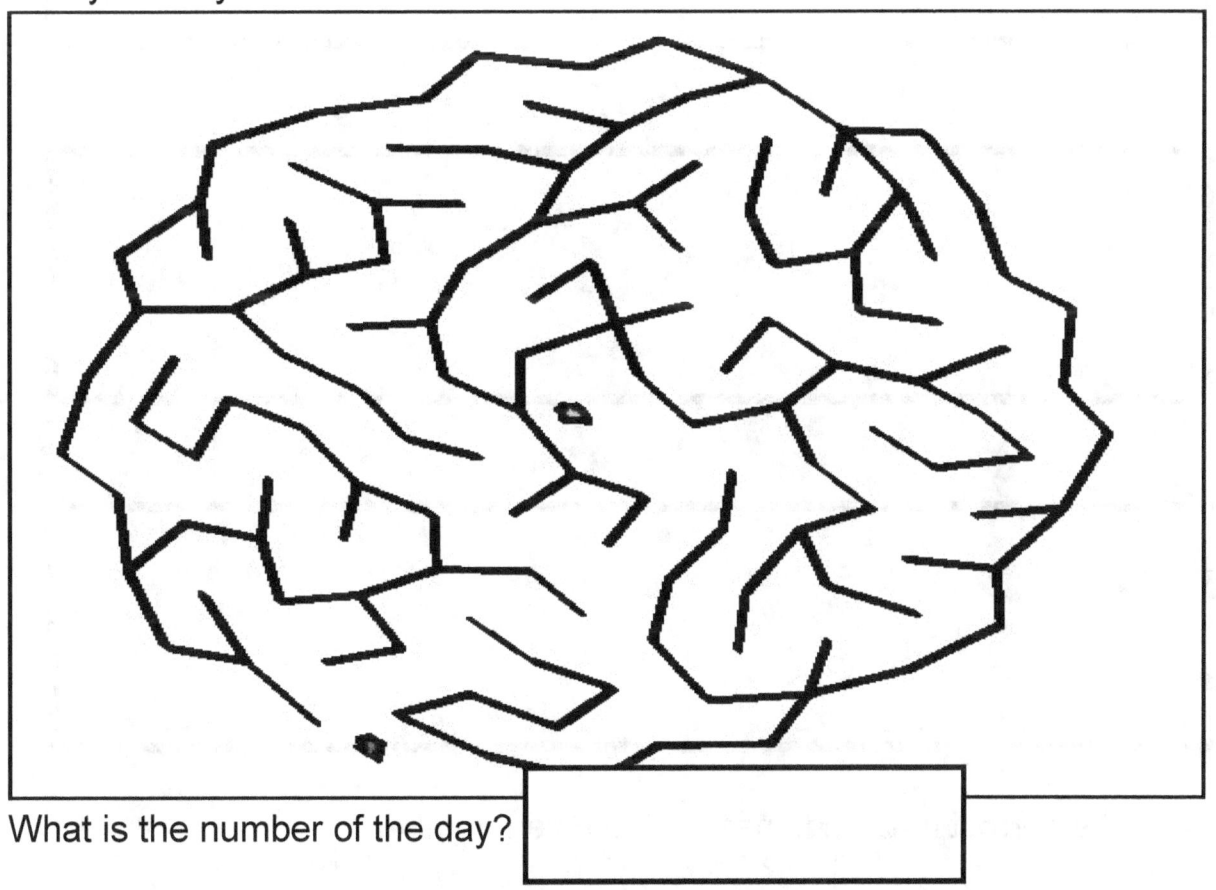

What is the number of the day? [ ]

Write the number using words:

Write the number three different ways using sticks and circles.

(Remember a stick represents a ten and a circle represents a one. If you know how to use a square block to represent a hundred, you may do so.)

Fill in the correct tens and ones for the given numbers.

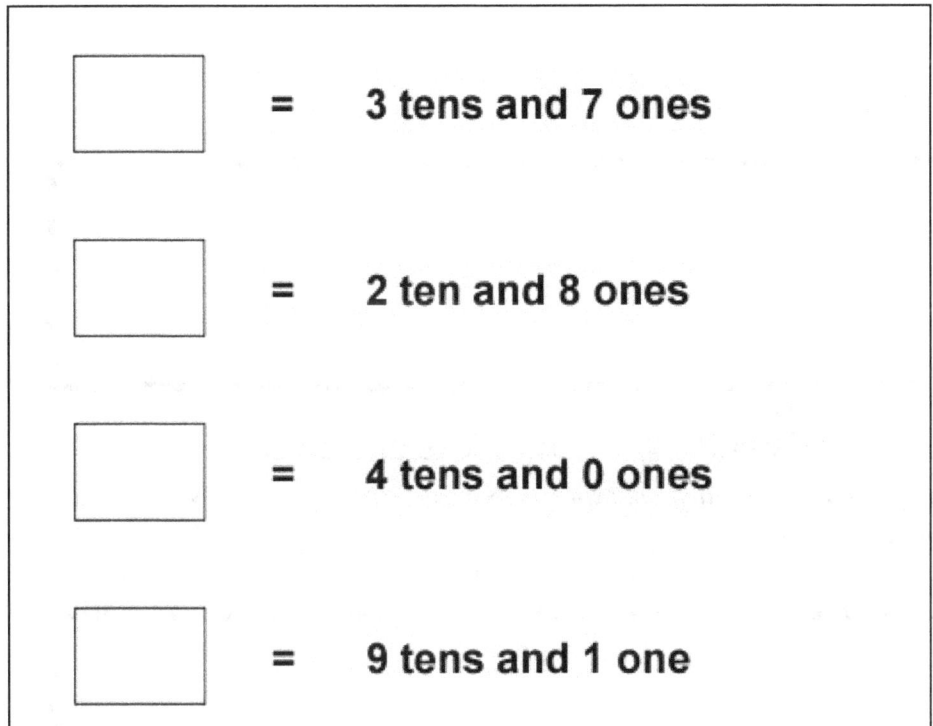

	=	3 tens and 7 ones
	=	2 ten and 8 ones
	=	4 tens and 0 ones
	=	9 tens and 1 one

Find your way out of the maze.

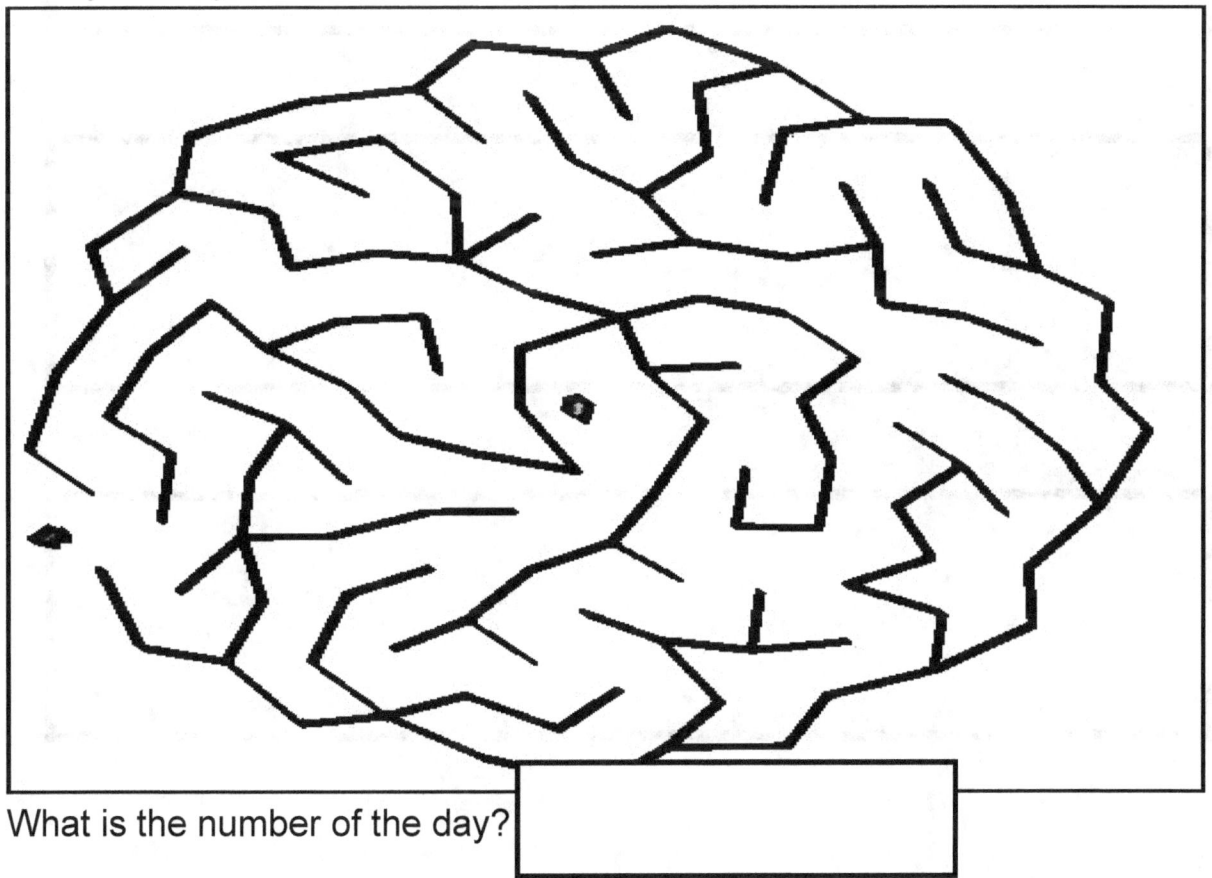

What is the number of the day?

Write the number using words:

Write the number three different ways using sticks and circles.
(Remember a stick represents a ten and a circle represents a one. If you know how to use a square block to represent a hundred, you may do so.)

Determine the place value of the underlined digit.

Example: 5̲3 = ___5 tens___

1) 8̲4 = _____

2) 2̲8 = _____

3) 2̲4 = _____

4) 2̲ = _____

Find your way out of the maze.

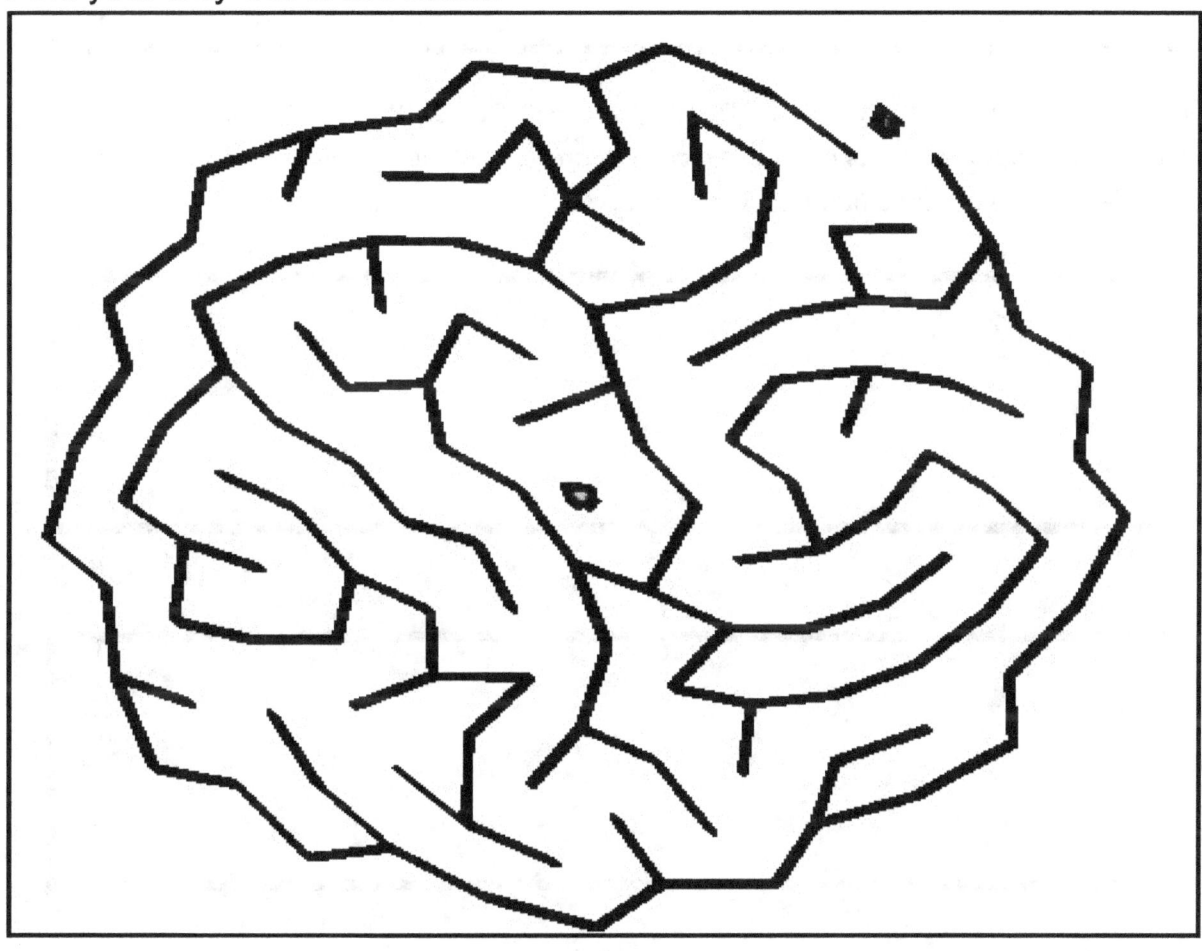

What is the number of the day?

Write the number using words:

Write the number three different ways using sticks and circles.
(Remember a stick represents a ten and a circle represents a one. If you know how to use a square block to represent a hundred, you may do so.)

Determine the place value of the underlined digit.

Example: 5̲3 = ___5 tens___

1) 8̲4 = _____

2) 4̲8 = _____

3) 4̲4̲ = _____

4) 9̲ = _____

Find your way out of the maze.

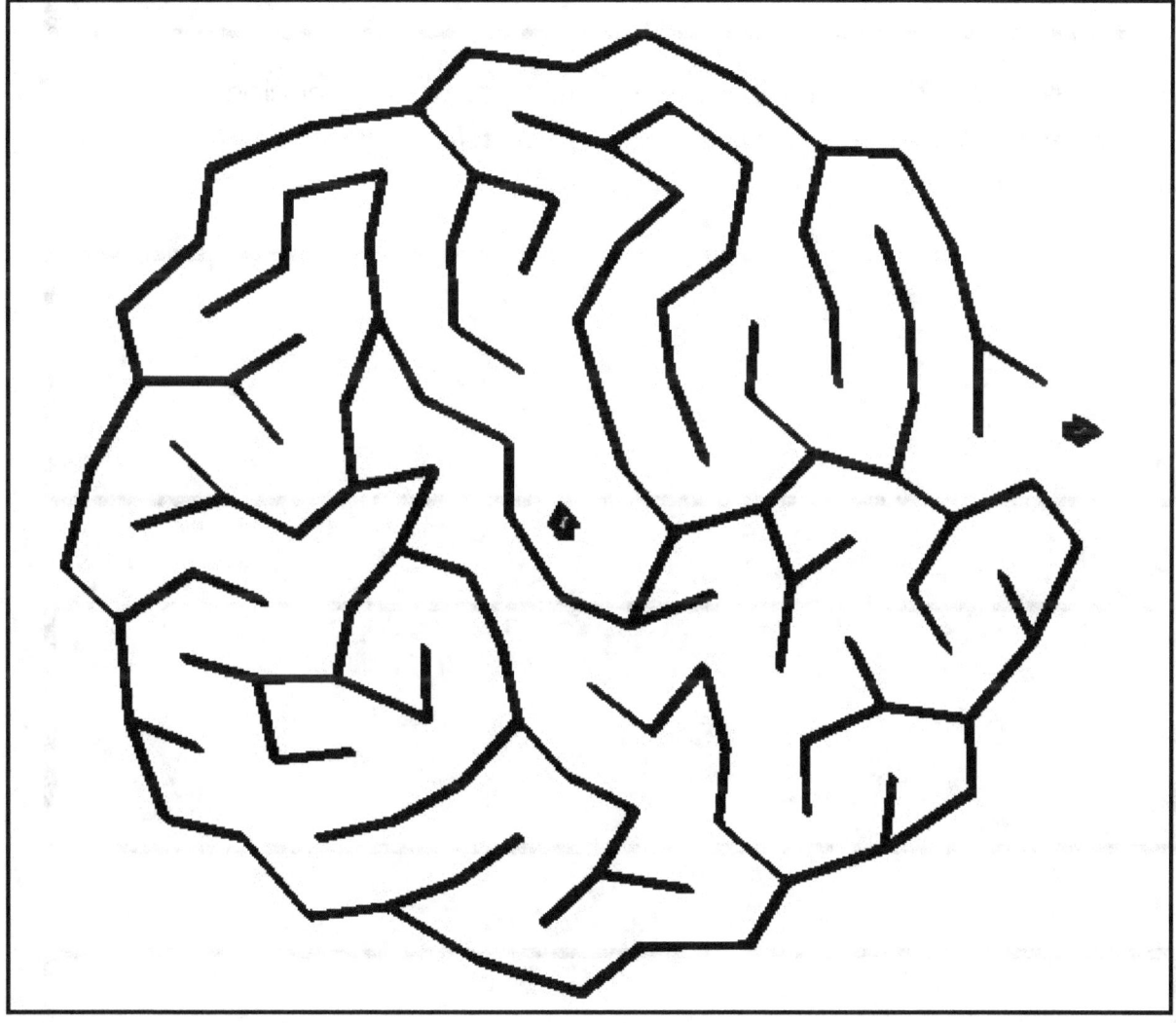

25

What is the number of the day?

Write the number using words:

Write the number three different ways using sticks and circles.
(Remember a stick represents a ten and a circle represents a one. If you know how to use a square block to represent a hundred, you may do so.)

Determine the place value of the underlined digit.

Example: 5̲3 = ___5 tens___

1) 6̲0 = _____   2) 4̲ = _____

3) 6̲ = _____   4) 2̲ = _____

Find your way out of the maze.

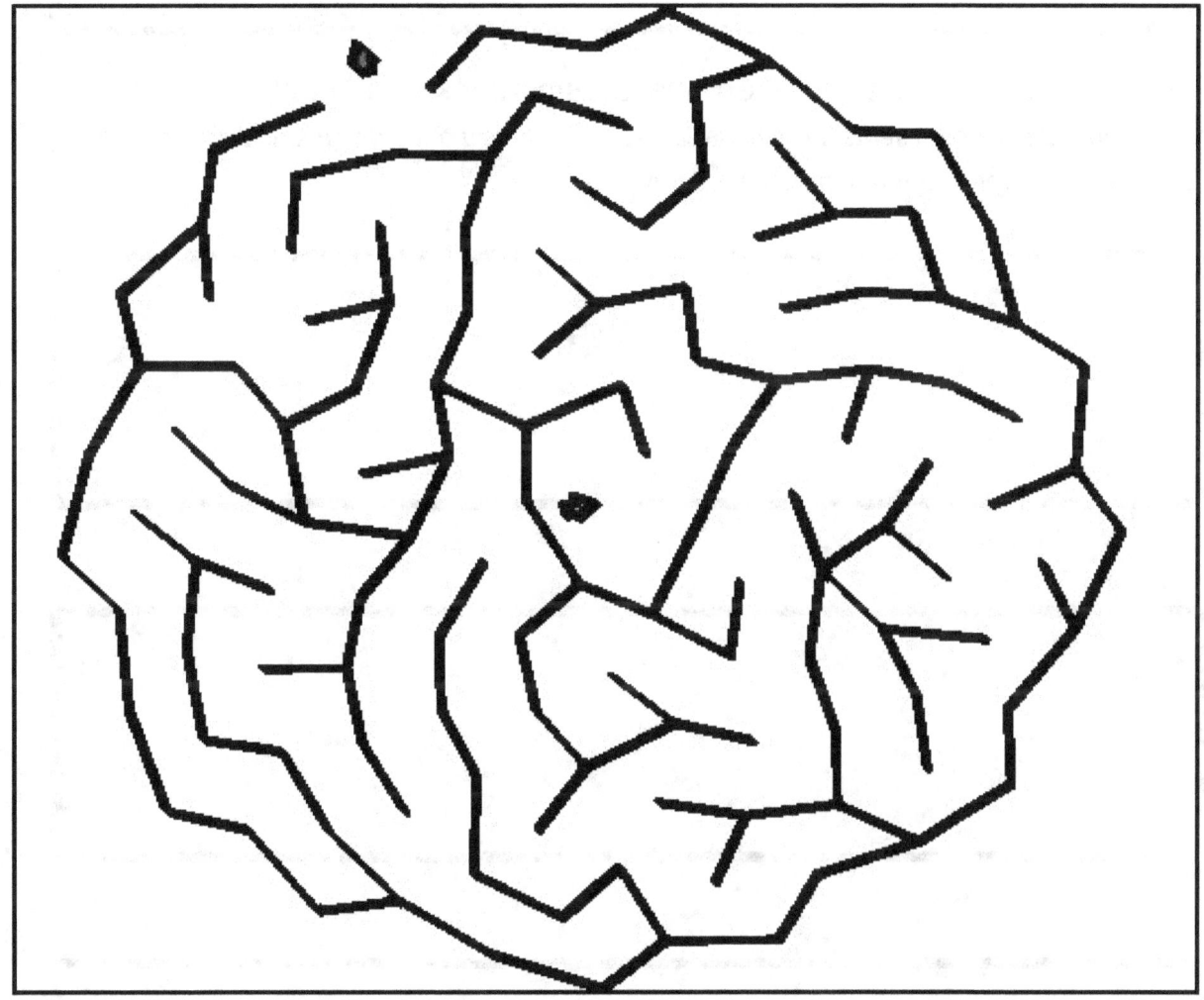

27

What is the number of the day?

Write the number using words:

Write the number three different ways using sticks and circles.
(Remember a stick represents a ten and a circle represents a one. If you know how to use a square block to represent a hundred, you may do so.)

Determine the place value of the underlined digit.

Example: 5̲3 = ___5 tens

1) 8̲3 = _____

2) 2̲ = _____

3) 4̲ = _____

4) 34̲ = _____

Find your way out of the maze.

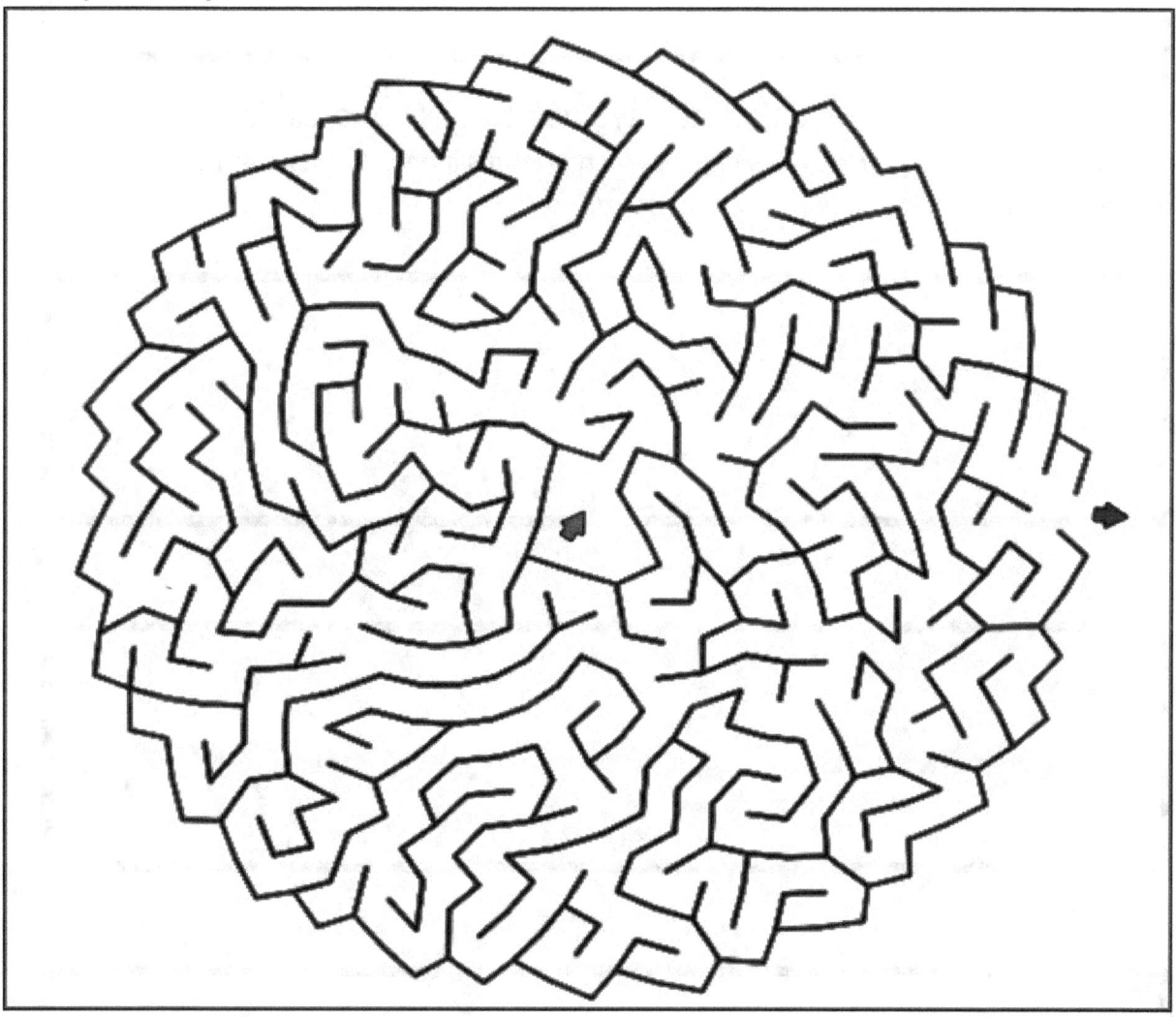

What is the number of the day?

Write the number using words:

Write the number three different ways using sticks and circles.
(Remember a stick represents a ten and a circle represents a one. If you know how to use a square block to represent a hundred, you may do so.)

Determine the place value of the underlined digit.

Example: 5̲3 = ___5 tens___

1) 6̲4̲ = _____

2) 9̲5 = _____

3) 5̲ = _____

4) 1̲0 = _____

Find your way out of the maze.

31

What is the number of the day?

Write the number using words:

Write the number three different ways using sticks and circles.
(Remember a stick represents a ten and a circle represents a one. If you know how to use a square block to represent a hundred, you may do so.)

Find the sum.

1) 30 + 4 = _____

2) 80 + 3 = _____

3) 50 + 6 = _____

4) 60 + 8 = _____

Find your way out of the maze.

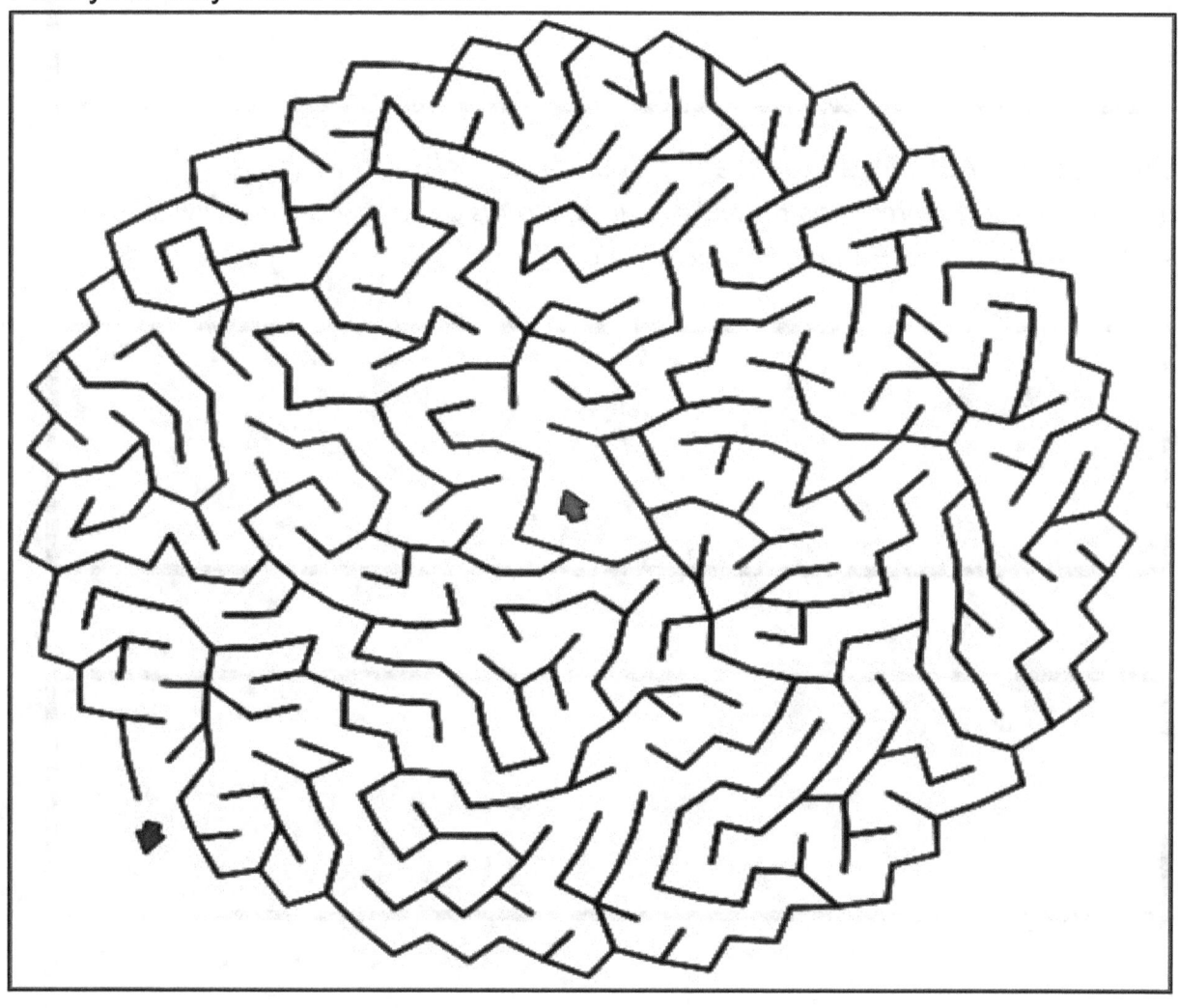

What is the number of the day?

Write the number using words:

Write the number three different ways using sticks and circles.
(Remember a stick represents a ten and a circle represents a one. If you know how to use a square block to represent a hundred, you may do so.)

Find the sum.

1) 90 + 8 = _____

2) 80 + 7 = _____

3) 80 + 3 = _____

4) 50 + 7 = _____

Find your way out of the maze.

What is the number of the day?

Write the number using words:

Write the number three different ways using sticks and circles.
(Remember a stick represents a ten and a circle represents a one. If you know how to use a square block to represent a hundred, you may do so.)

Find the sum.

1) 70 + 9 = _____

2) 60 + 9 = _____

3) 70 + 6 = _____

4) 80 + 6 = _____

Find your way out of the maze.

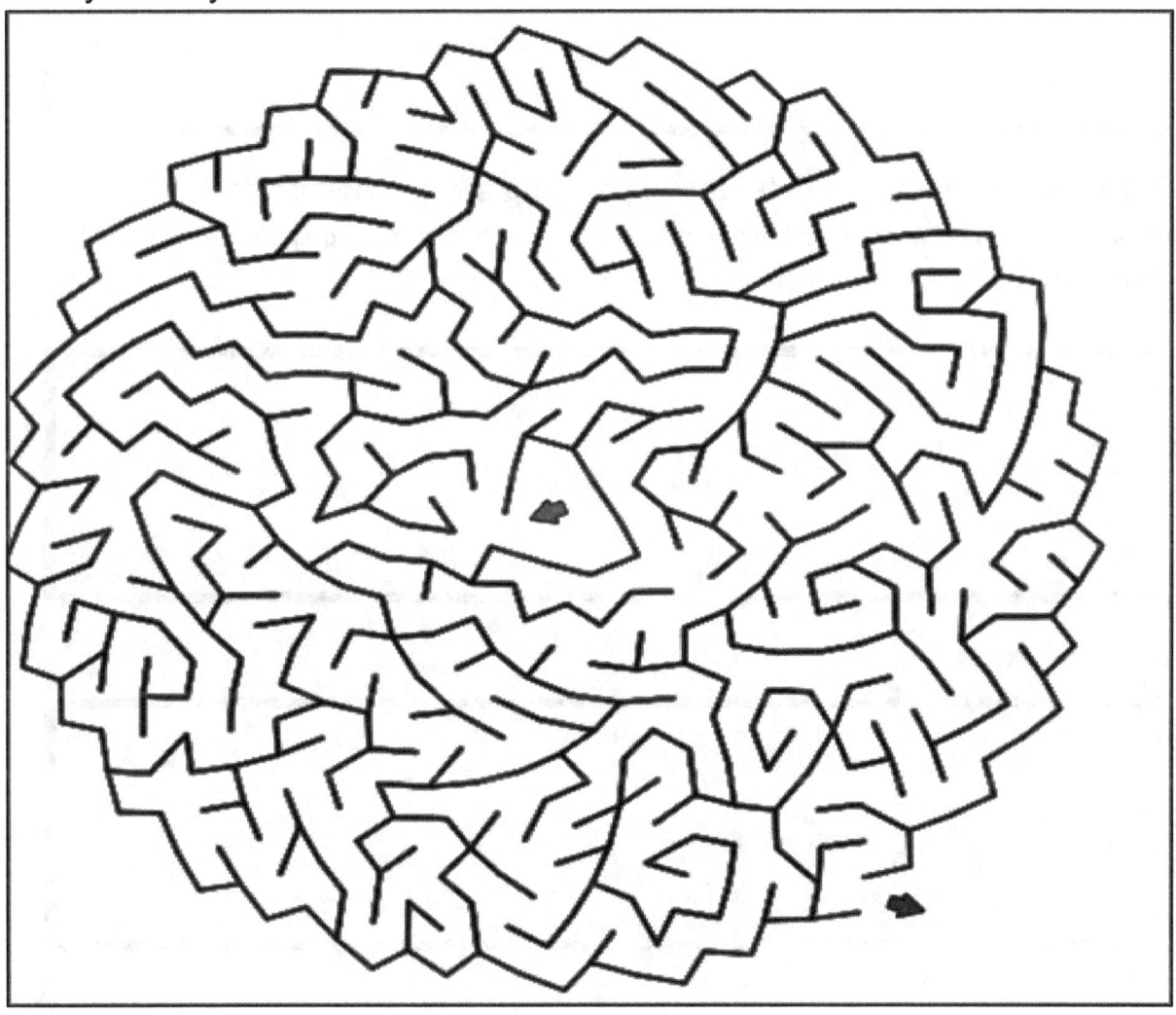

What is the number of the day?

Write the number using words:

Write the number three different ways using sticks and circles.
(Remember a stick represents a ten and a circle represents a one. If you know how to use a square block to represent a hundred, you may do so.)

Fill in the missing numbers.

1) Count by 2 from 7 to 23

			13		17			23

2) Count by 1 from 5 to 13

		7	8			11		

3) Count by 2 from 1 to 17

				9	11			17

4) Count by 1 from 3 to 11

	4		6	7				

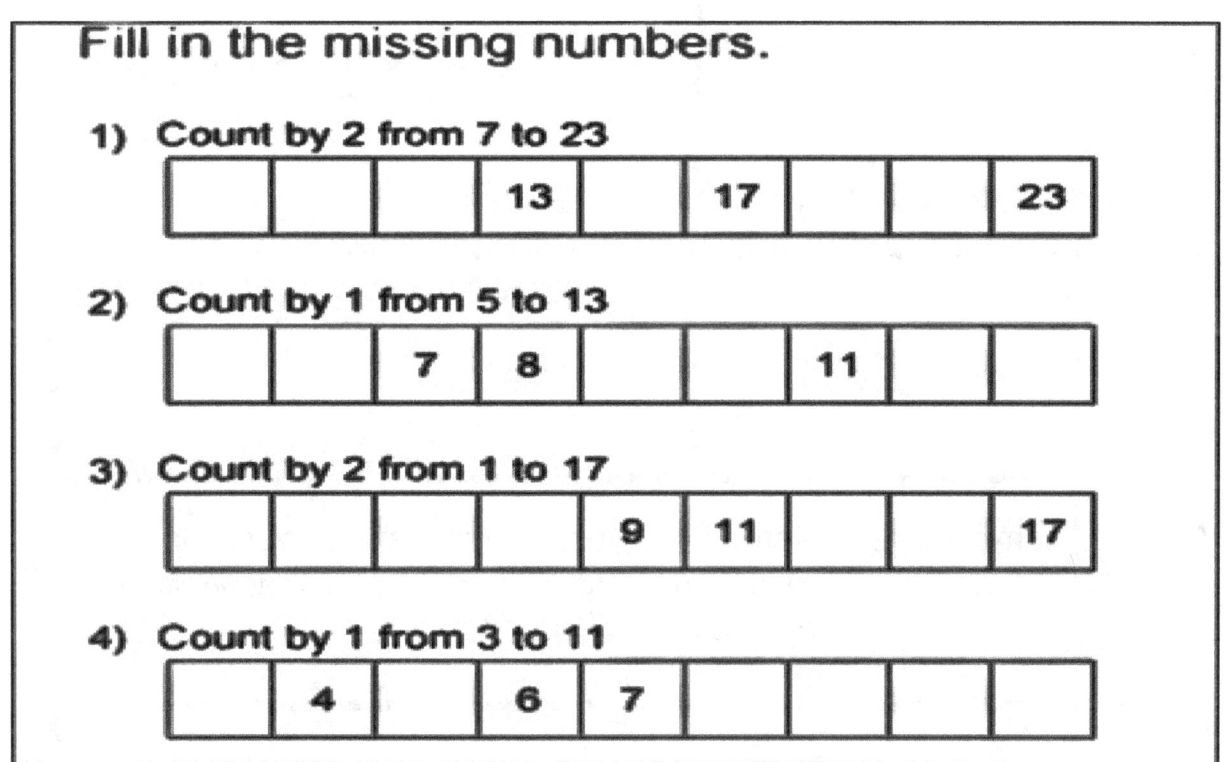

Find your way out of the maze.

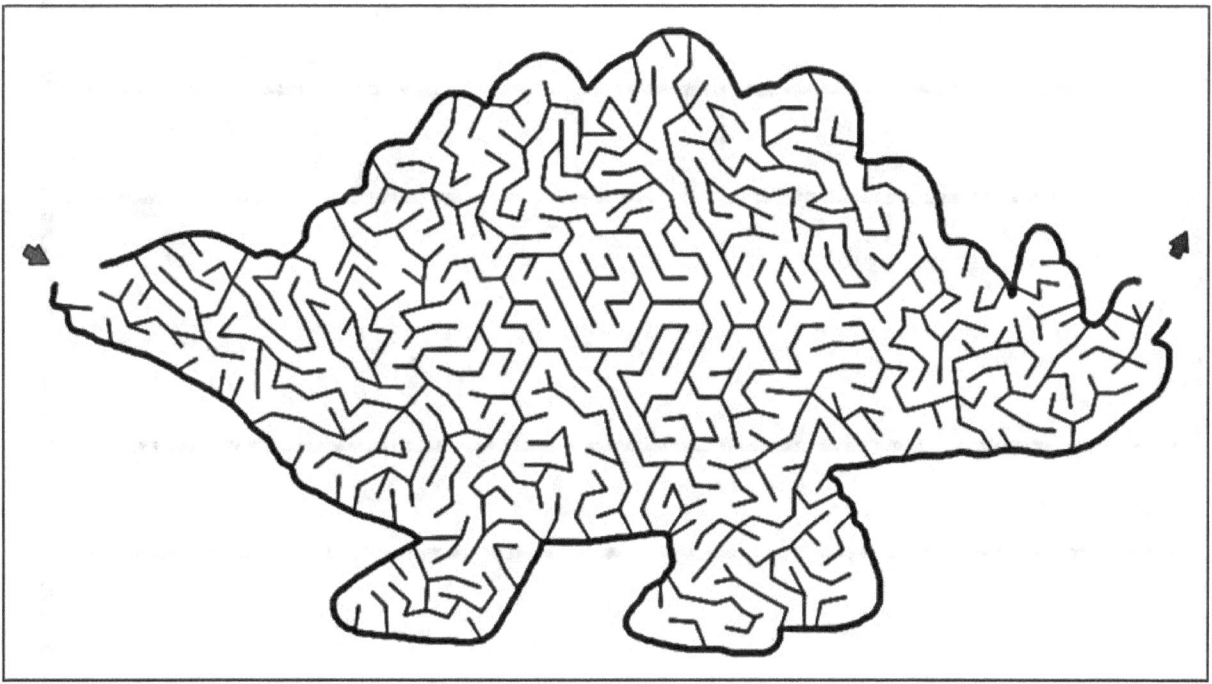

What is the number of the day?

Write the number using words:

Write the number three different ways using sticks and circles.
(Remember a stick represents a ten and a circle represents a one. If you know how to use a square block to represent a hundred, you may do so.)

Fill in the missing numbers.

1) Count by 1 from 2 to 10

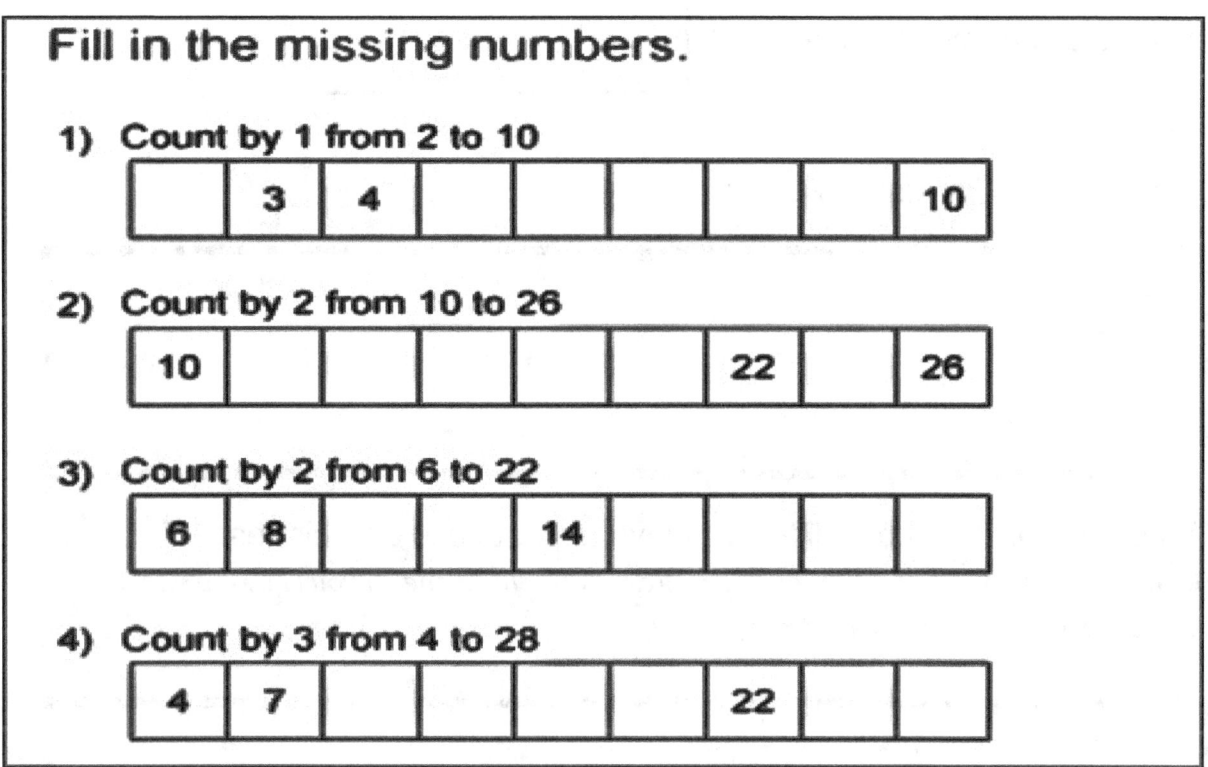

	3	4						10

2) Count by 2 from 10 to 26

10					22		26

3) Count by 2 from 6 to 22

6	8			14				

4) Count by 3 from 4 to 28

4	7				22		

Find your way out of the maze.

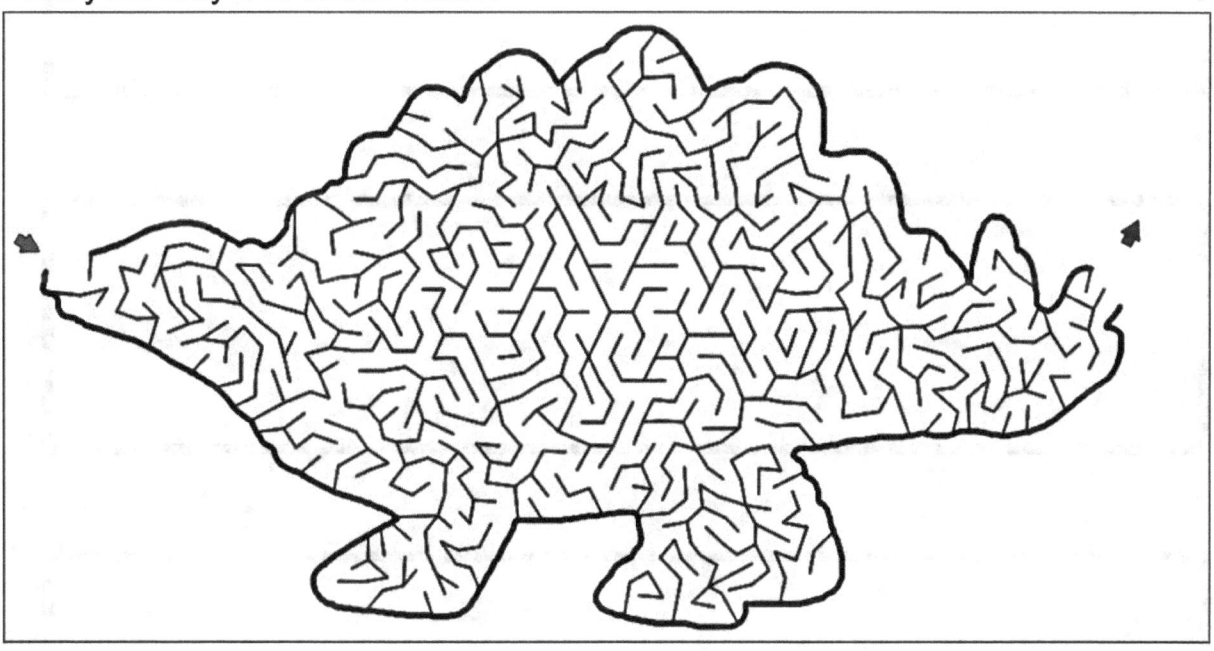

What is the number of the day?

Write the number using words:

Write the number three different ways using sticks and circles.
(Remember a stick represents a ten and a circle represents a one. If you know how to use a square block to represent a hundred, you may do so.)

Fill in the missing numbers.

1) Count by 1 from 48 to 41

	47				43		

2) Count by 3 from 43 to 22

			34			25	

3) Count by 2 from 47 to 33

			41			35	

4) Count by 1 from 33 to 26

33						27	

Find your way out of the maze.

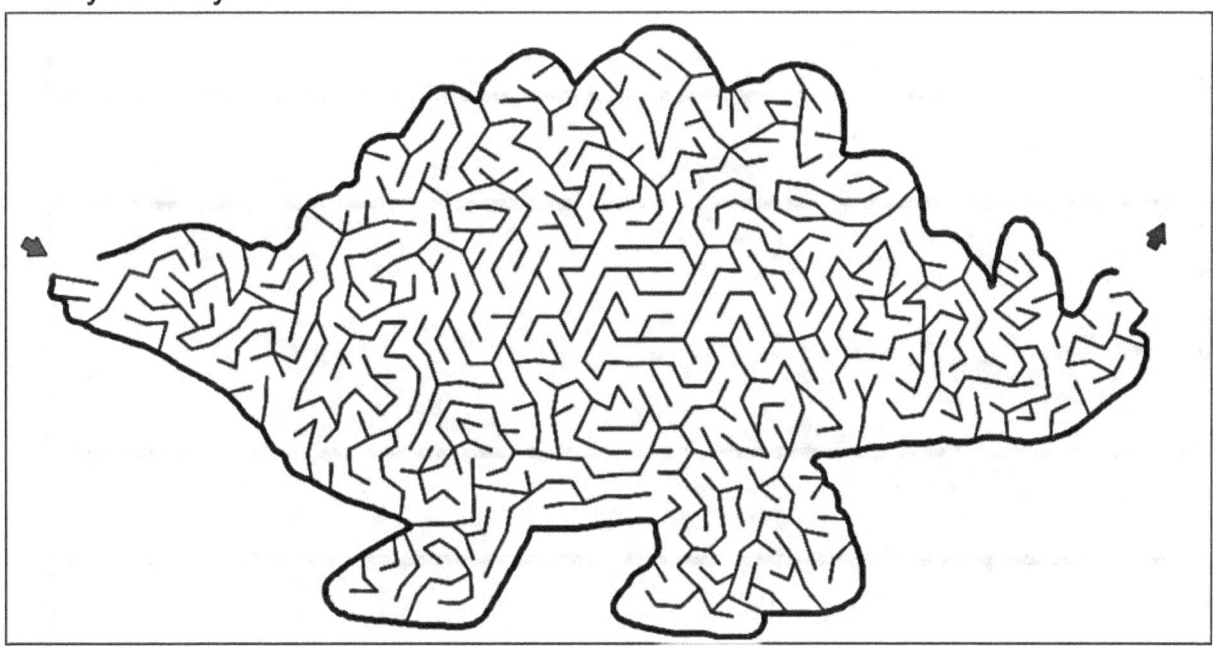

What is the number of the day?

Write the number using words:

Write the number three different ways using sticks and circles.
(Remember a stick represents a ten and a circle represents a one. If you know how to use a square block to represent a hundred, you may do so.)

Fill in the missing numbers.

1) Count by 1 from 37 to 30

				33			30

2) Count by 1 from 33 to 26

			30				26

3) Count by 3 from 39 to 18

		33	30				

4) Count by 2 from 23 to 9

			17		13		

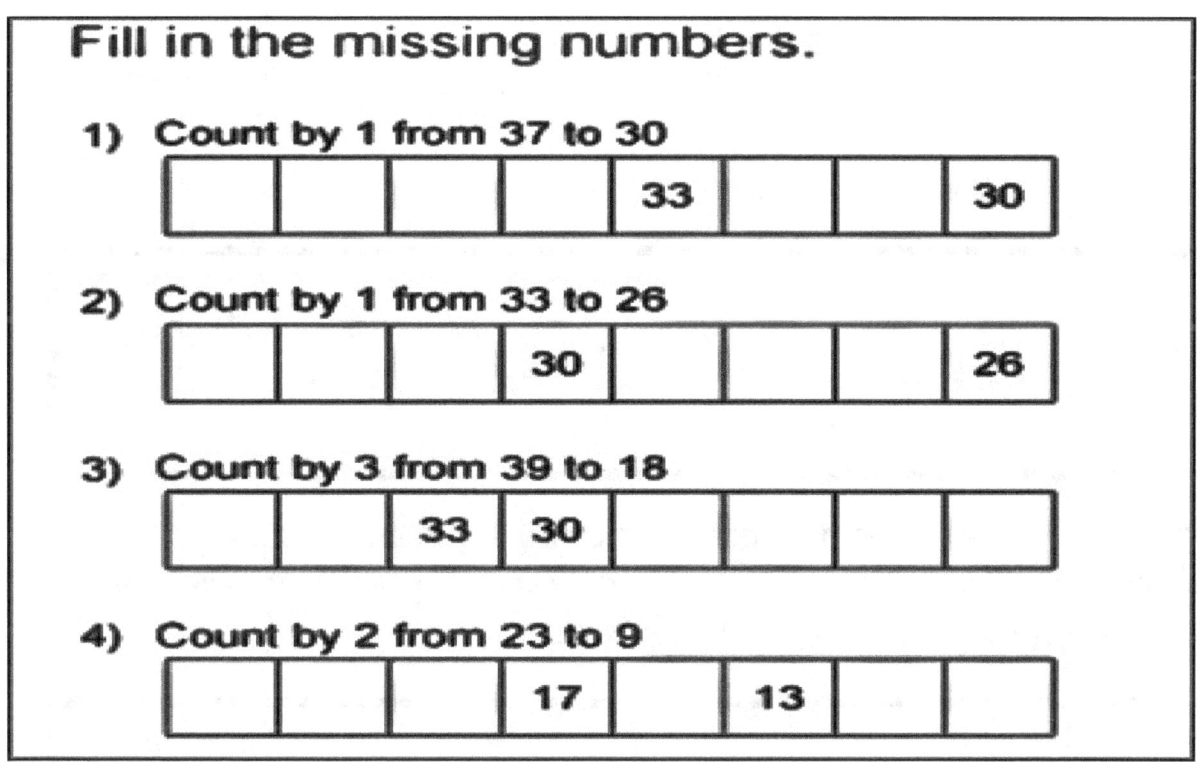

Find your way out of the maze.

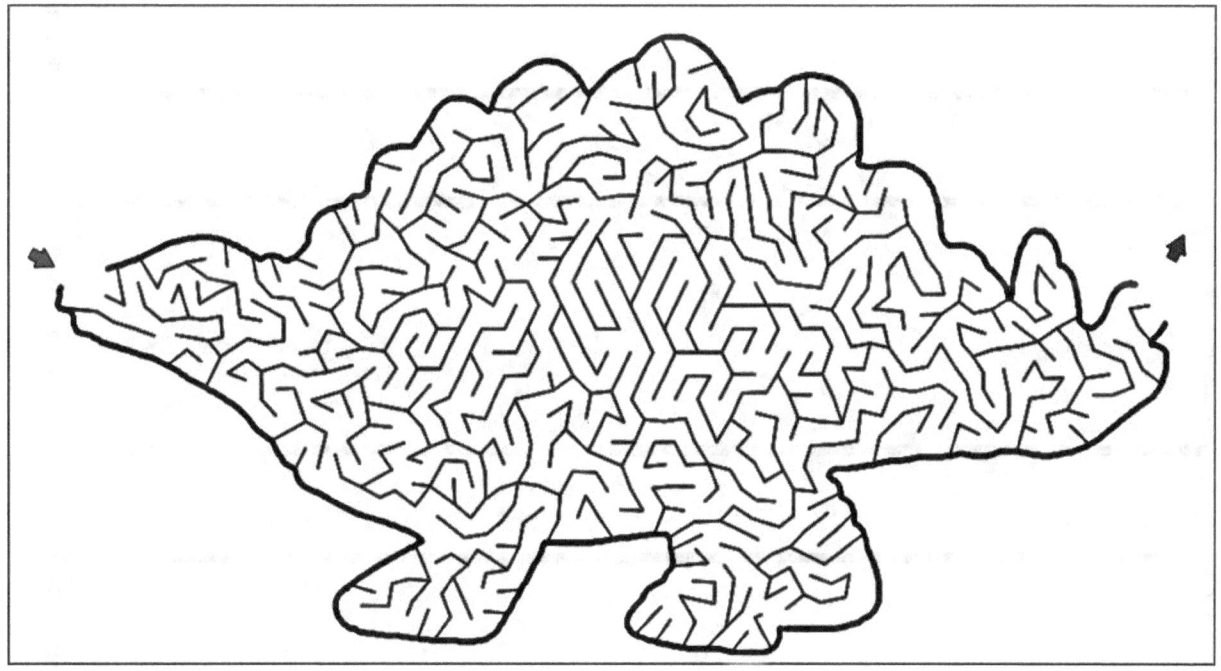

What is the number of the day?

Write the number using words:

Write the number three different ways using sticks and circles.
(Remember a stick represents a ten and a circle represents a one. If you know how to use a square block to represent a hundred, you may do so.)

Fill in the missing numbers.

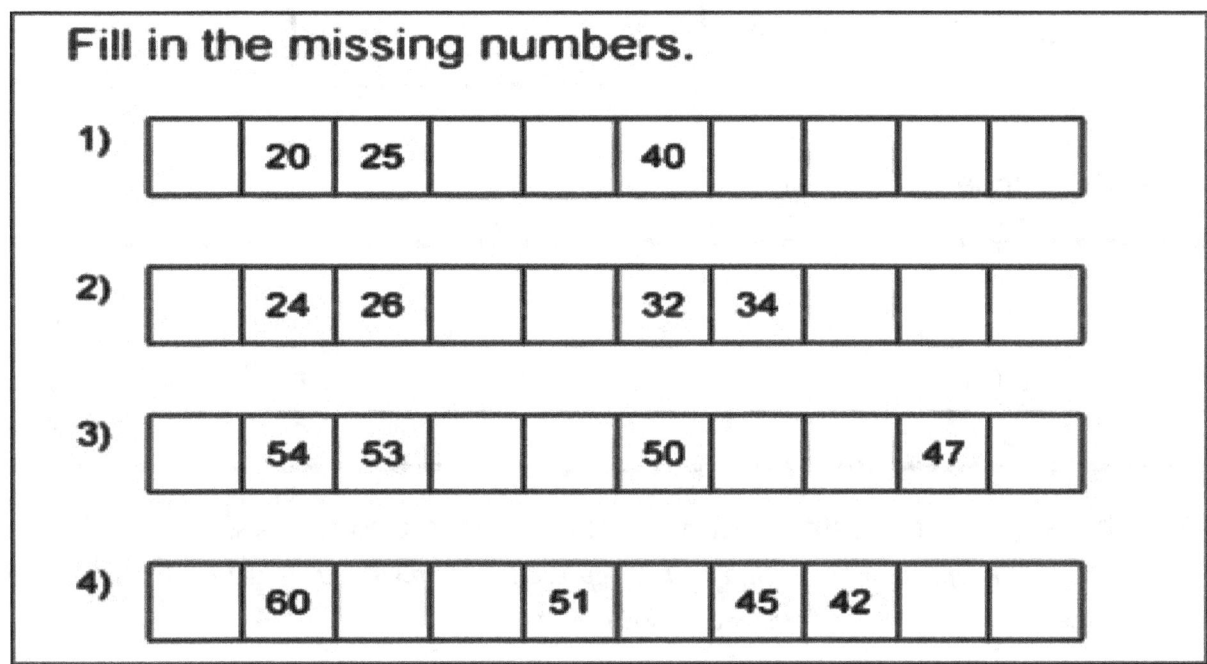

1) | | 20 | 25 | | | 40 | | | | |

2) | | 24 | 26 | | | 32 | 34 | | | |

3) | | 54 | 53 | | | 50 | | | 47 | |

4) | | 60 | | | 51 | | 45 | 42 | | |

Find your way out of the maze.

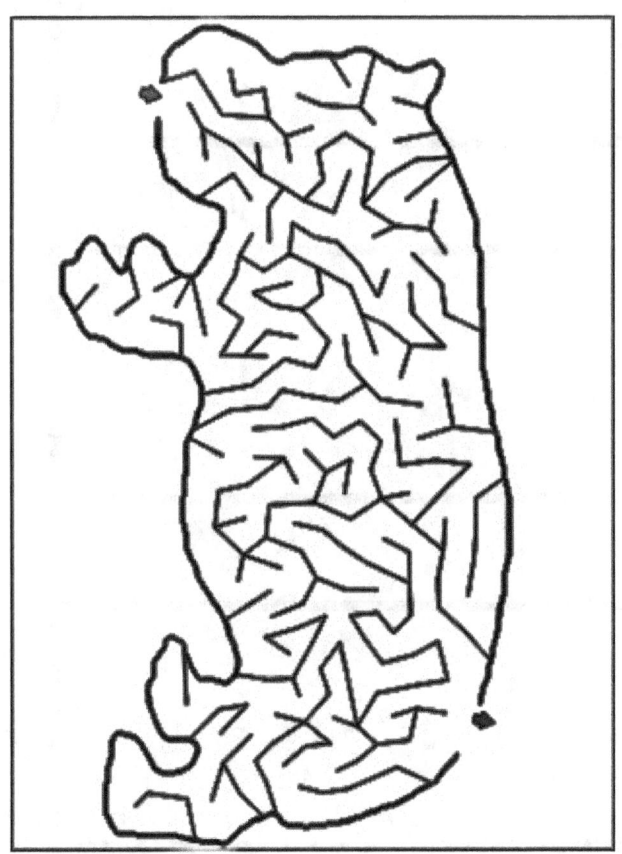

What is the number of the day?

Write the number using words:

Write the number three different ways using sticks and circles.
(Remember a stick represents a ten and a circle represents a one. If you know how to use a square block to represent a hundred, you may do so.)

Fill in the missing numbers.

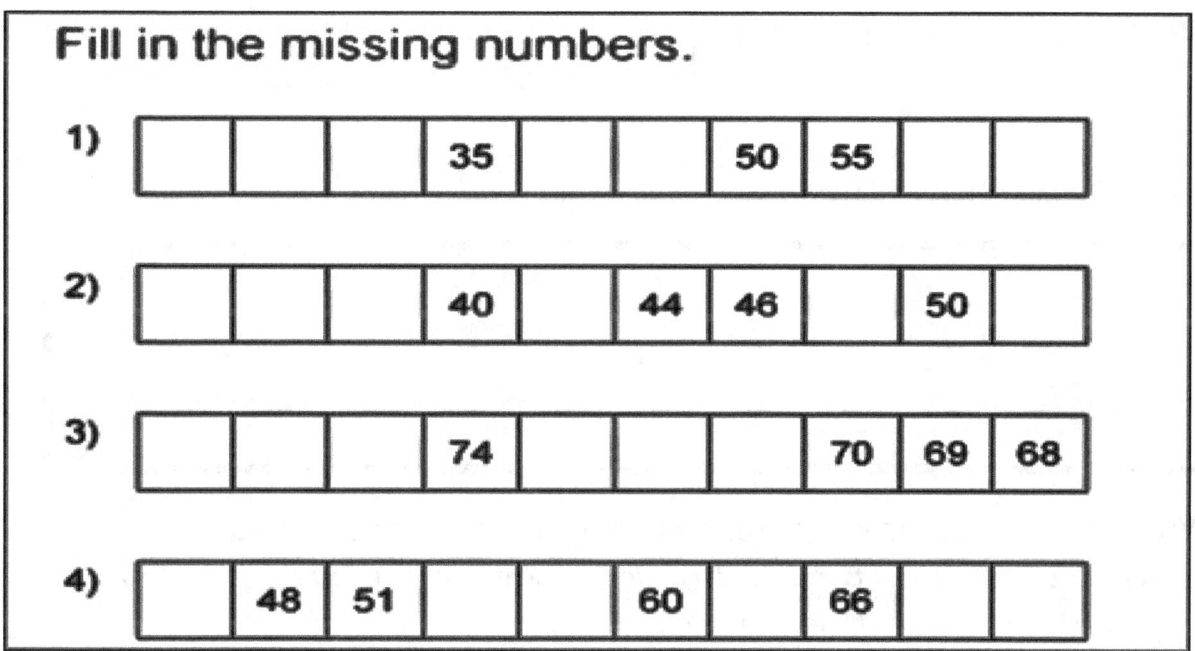

1)

			35			50	55		

2)

			40		44	46		50	

3)

			74				70	69	68

4)

	48	51			60		66		

Find your way out of the maze.

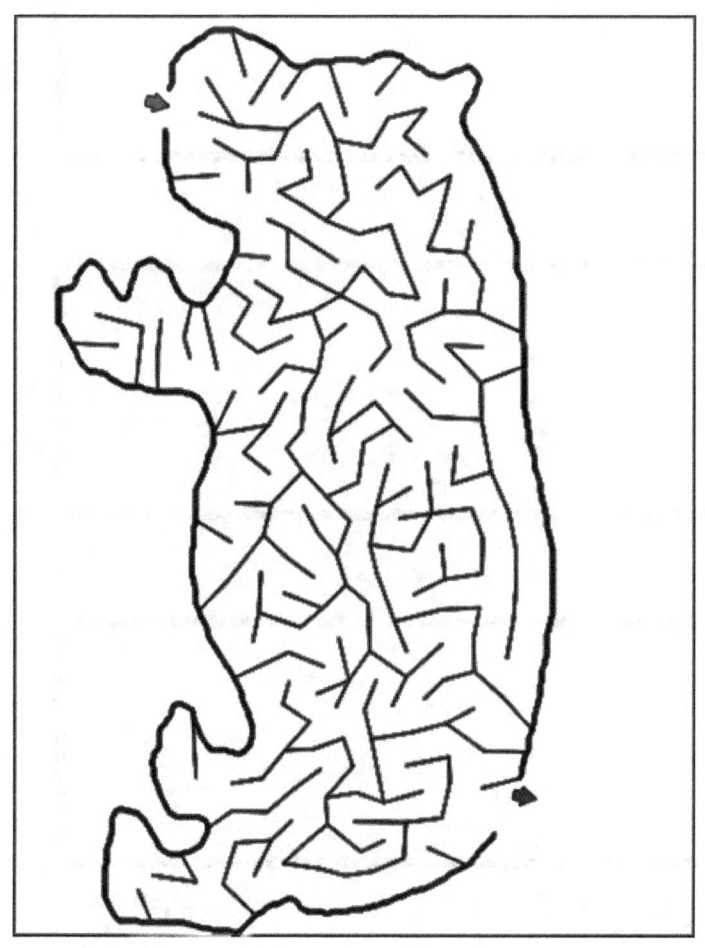

What is the number of the day?

Write the number using words:

Write the number three different ways using sticks and circles.
(Remember a stick represents a ten and a circle represents a one. If you know how to use a square block to represent a hundred, you may do so.)

Fill in the missing numbers.

1)

		74					64	62	

2)

2				6		8	9		

3)

60	55					25		15

4)

	51	54		60	63				

Find your way out of the maze.

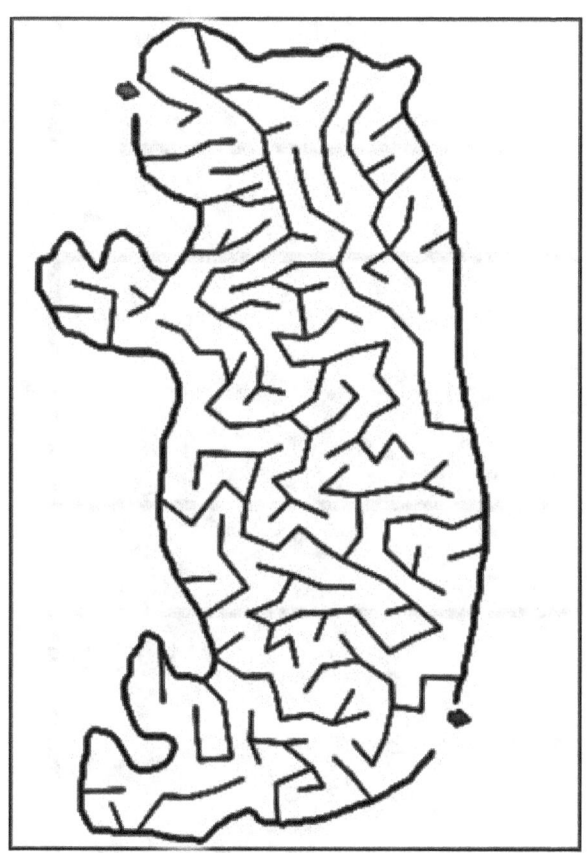

What is the number of the day?

Write the number using words:

Write the number three different ways using sticks and circles.
(Remember a stick represents a ten and a circle represents a one. If you know how to use a square block to represent a hundred, you may do so.)

Fill in the missing numbers.

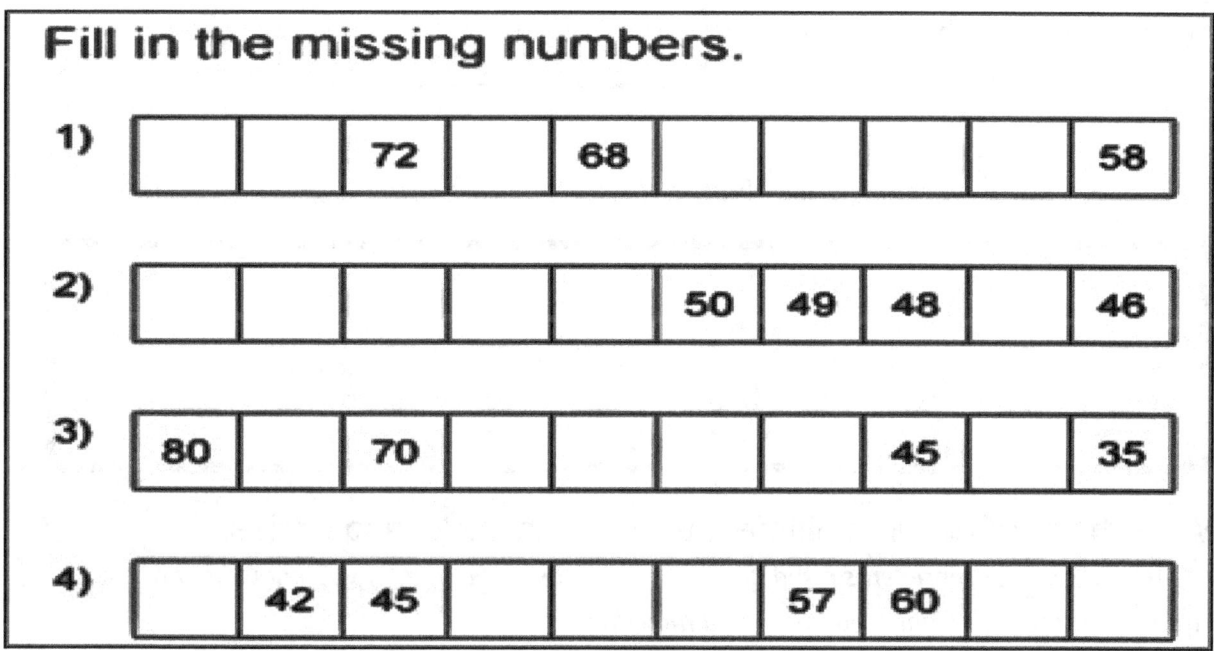

1)

		72		68					58

2)

					50	49	48		46

3)

80		70					45		35

4)

	42	45				57	60		

Find your way out of the maze.

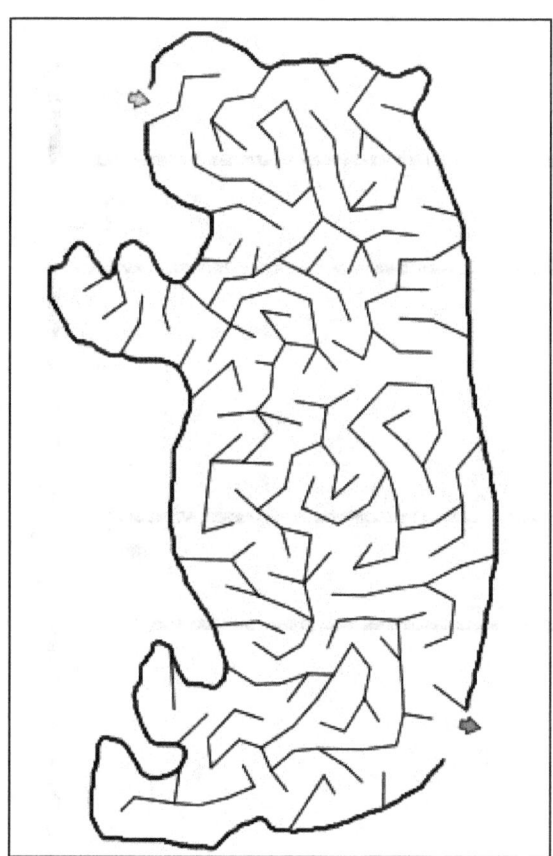

What is the number of the day?

Write the number using words:

Write the number three different ways using sticks and circles.
(Remember a stick represents a ten and a circle represents a one. If you know how to use a square block to represent a hundred, you may do so.)

Fill in the empty blanks. Write a rule to represent the relationship between input and output.

1)

Input	Output
2	11
5	14
8	
6	

_____

2)

Input	Output
1	10
8	17
9	
10	

_____

3)

Input	Output
1	5
8	12
5	
10	

_____

Find your way out of the maze.

What is the number of the day?

Write the number using words:

Write the number three different ways using sticks and circles.

(Remember a stick represents a ten and a circle represents a one. If you know how to use a square block to represent a hundred, you may do so.)

Fill in the empty blanks. Write a rule to represent the relationship between input and output.

1)

Input	Output
12	18
11	17
6	
7	

_____

2)

Input	Output
17	37
19	39
16	
13	

_____

3)

Input	Output
16	17
12	13
7	
8	

_____

Find your way out of the maze.

Sparky can't find his bone. Will you show him the way?

What is the number of the day?

Write the number using words:

Write the number three different ways using sticks and circles.
(Remember a stick represents a ten and a circle represents a one. If you know how to use a square block to represent a hundred, you may do so.)

What number should be added to the first number to make 8?

1) $4 + \underline{} = 8$

2) $5 + \underline{} = 8$

3) $2 + \underline{} = 8$

4) $7 + \underline{} = 8$

5) $6 + \underline{} = 8$

6) $3 + \underline{} = 8$

7) $8 + \underline{} = 8$

8) $1 + \underline{} = 8$

9) $4 + \underline{} = 8$

10) $8 + \underline{} = 8$

Find your way out of the maze.

Help the bat find the way to the belfry.

What is the number of the day?

Write the number using words:

Write the number three different ways using sticks and circles.
(Remember a stick represents a ten and a circle represents a one. If you know how to use a square block to represent a hundred, you may do so.)

What number should be added to the first number to make 9?

1) $2 + \underline{} = 9$

2) $3 + \underline{} = 9$

3) $4 + \underline{} = 9$

4) $8 + \underline{} = 9$

5) $5 + \underline{} = 9$

6) $1 + \underline{} = 9$

7) $7 + \underline{} = 9$

8) $6 + \underline{} = 9$

9) $4 + \underline{} = 9$

10) $4 + \underline{} = 9$

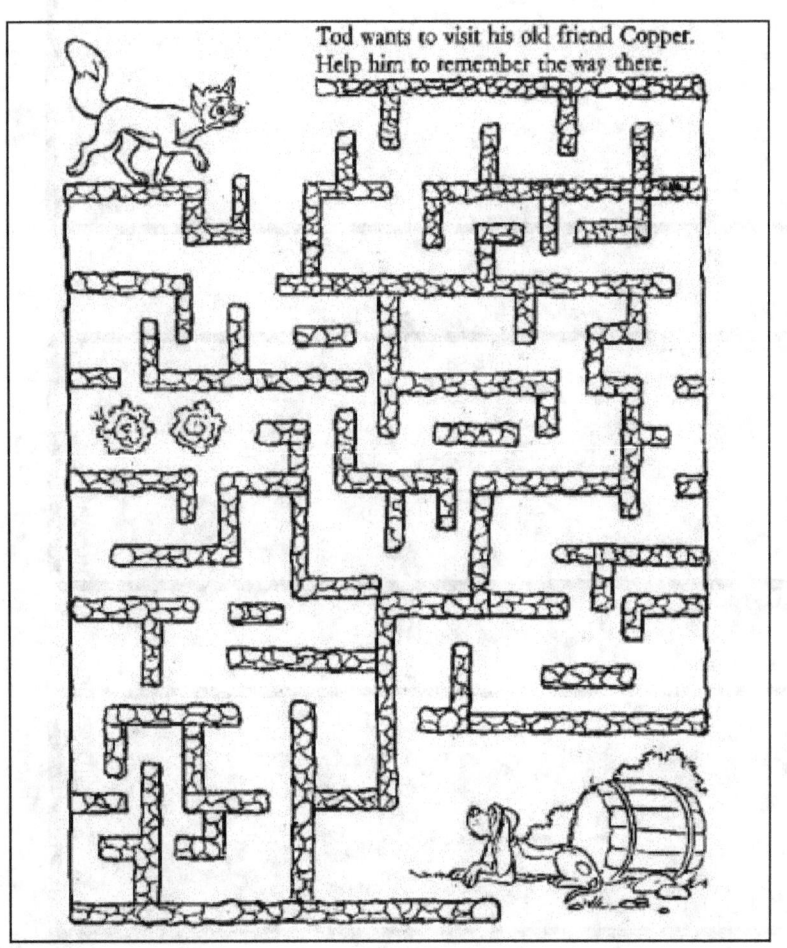

Tod wants to visit his old friend Copper.
Help him to remember the way there.

What is the number of the day?

Write the number using words:

Write the number three different ways using sticks and circles.
(Remember a stick represents a ten and a circle represents a one. If you know how to use a square block to represent a hundred, you may do so.)

What number should be added to the first number to make 10?

1) 7 + ___ = 10

2) 4 + ___ = 10

3) 3 + ___ = 10

4) 8 + ___ = 10

5) 5 + ___ = 10

6) 6 + ___ = 10

7) 1 + ___ = 10

8) 2 + ___ = 10

9) 6 + ___ = 10

10) 8 + ___ = 10

63

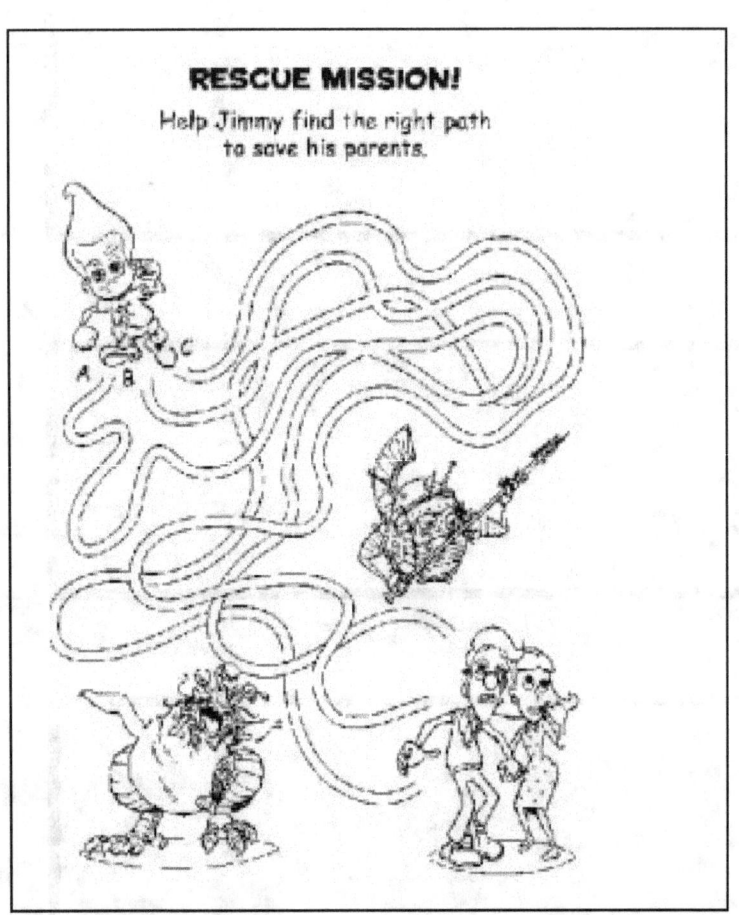

RESCUE MISSION!

Help Jimmy find the right path
to save his parents.

What is the number of the day?

Write the number using words:

Write the number three different ways using sticks and circles.
(Remember a stick represents a ten and a circle represents a one. If you know how to use a square block to represent a hundred, you may do so.)

Write the sum in the box provided.

$5 + 5 = \boxed{}$	$6 + 6 = \boxed{}$	$4 + 4 = \boxed{}$
$2 + 2 = \boxed{}$	$1 + 1 = \boxed{}$	$9 + 9 = \boxed{}$
$3 + 3 = \boxed{}$	$8 + 8 = \boxed{}$	$5 + 5 = \boxed{}$

What is the number of the day?

Write the number using words:

Write the number three different ways using sticks and circles.
(Remember a stick represents a ten and a circle represents a one. If you know how to use a square block to represent a hundred, you may do so.)

Write the sum in the box provided.

8 + 8 = ☐   2 + 2 = ☐   9 + 9 = ☐

1 + 1 = ☐   4 + 4 = ☐   5 + 5 = ☐

9 + 9 = ☐   0 + 0 = ☐   7 + 7 = ☐

Help the bear find the music.

What is the number of the day?

Write the number using words:

Write the number three different ways using sticks and circles.
(Remember a stick represents a ten and a circle represents a one. If you know how to use a square block to represent a hundred, you may do so.)

Use an adding doubles strategy to find each sum.

Example: 2 + 3 = 2 + 2 + 1 = 4 + 1 = 5

4 + 5 = ☐   1 + 2 = ☐   8 + 9 = ☐

6 + 7 = ☐   7 + 8 = ☐   5 + 6 = ☐

0 + 1 = ☐   3 + 4 = ☐   2 + 3 = ☐

CASTLE MAZE

Start

Finish

What is the number of the day?

Write the number using words:

Write the number three different ways using sticks and circles.
(Remember a stick represents a ten and a circle represents a one. If you know how to use a square block to represent a hundred, you may do so.)

Use an adding doubles strategy to find each sum.

Example: 2 + 3 = 2 + 2 + 1 = 4 + 1 = 5

1 + 2 = ☐   3 + 4 = ☐   6 + 7 = ☐

7 + 8 = ☐   5 + 6 = ☐   8 + 9 = ☐

2 + 3 = ☐   0 + 1 = ☐   6 + 7 = ☐

FINISH

See if you can start at the bottom and find your way to the top.

START

What is the number of the day?

Write the number using words:

Write the number three different ways using sticks and circles.
(Remember a stick represents a ten and a circle represents a one. If you know how to use a square block to represent a hundred, you may do so.)

Use an adding doubles strategy to find each sum.

Example: 2 + 3 = 2 + 2 + 1 = 4 + 1 = 5

$6 + 7 =$ ☐ $2 + 3 =$ ☐ $5 + 6 =$ ☐

$8 + 9 =$ ☐ $0 + 1 =$ ☐ $4 + 5 =$ ☐

$5 + 6 =$ ☐ $3 + 4 =$ ☐ $1 + 2 =$ ☐

Will you help Jessica find the way to her duck?

Start

Finish

74

What is the number of the day?

Write the number using words:

Write the number three different ways using sticks and circles.
(Remember a stick represents a ten and a circle represents a one. If you know how to use a square block to represent a hundred, you may do so.)

Use an adding doubles strategy to find each sum.

Example: 2 + 3 = 2 + 2 + 1 = 4 + 1 = 5

7 + 8 = ☐   4+ 5 = ☐   2 + 3 = ☐

1 + 2 = ☐   0 + 1 = ☐   6 + 7 = ☐

3 + 4 = ☐   5 + 6 = ☐   8 + 9 = ☐

Show this hen the
way back to the barn.

What is the number of the day?

Write the number using words:

Write the number three different ways using sticks and circles.
(Remember a stick represents a ten and a circle represents a one. If you know how to use a square block to represent a hundred, you may do so.)

1. Add.

a.	b.	c.	d.
2 + 3 = _____	7 + 3 = _____	6 + 2 = _____	5 + 5 = _____
4 + 4 = _____	5 + 4 = _____	4 + 6 = _____	2 + 4 = _____
1 + 6 = _____	3 + 6 = _____	2 + 5 = _____	9 + 1 = _____
2 + 7 = _____	1 + 7 = _____	6 + 2 = _____	5 + 3 = _____

2. Subtract.

a.	b.	c.	d.
8 – 3 = _____	5 – 3 = _____	7 – 3 = _____	10 – 3 = _____
6 – 4 = _____	7 – 4 = _____	9 – 4 = _____	5 – 4 = _____
10 – 6 = _____	9 – 6 = _____	4 – 3 = _____	8 – 6 = _____
8 – 7 = _____	6 – 3 = _____	10 – 7 = _____	9 – 7 = _____

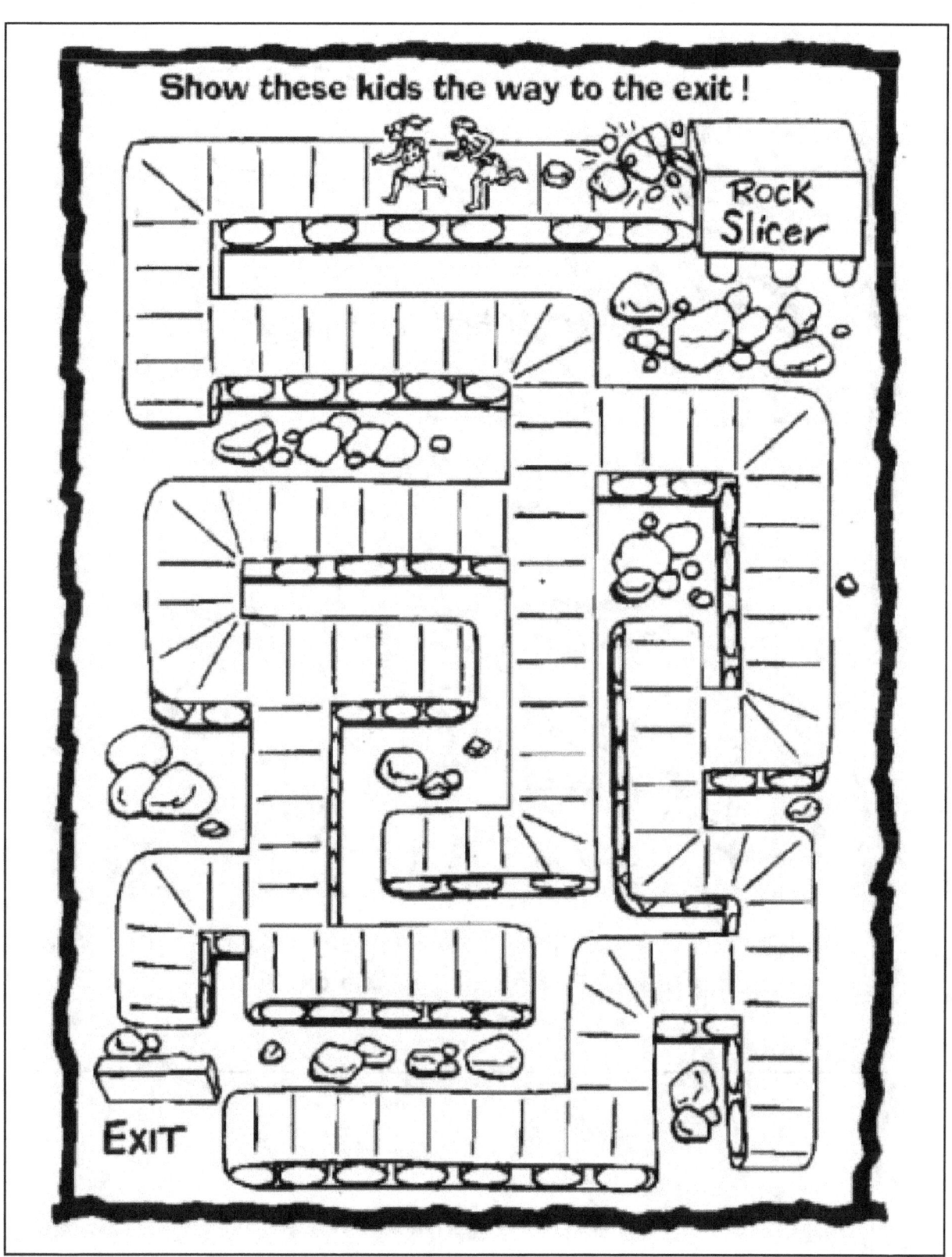

Finish

Start

Help the Knight find his way to the castle.

What is the number of the day?

Write the number using words:

Write the number three different ways using sticks and circles.
(Remember a stick represents a ten and a circle represents a one. If you know how to use a square block to represent a hundred, you may do so.)

3. Write a fact family to match the picture.

_____ + _____ = _____   _____ + _____ = _____

_____ - _____ = _____   _____ - _____ = _____

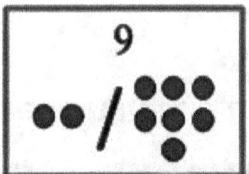

4. Find the missing numbers.

a. 2 + _____ = 7	b. 1 + _____ = 8	c. 4 + _____ = 6	d. _____ + 3 = 8
3 + _____ = 8	2 + _____ = 10	_____ + 3 = 9	_____ + 6 = 10

Place Value and Two-Digit Numbers

5. Fill in the missing parts.

a. 20 + 7 = _____	b. 6 + _____ = 56	c. 40 + _____ = 40
5 + 60 = _____	30 + _____ = 39	4 + _____ = 94

6. Put the numbers in order.

a. 16, 61, 26	b. 54, 14, 51
_____ < _____ < _____	_____ < _____ < _____

7. Compare the expressions and write <, >, or =.

a. 40 + 8 [ ] 4 + 80   b. 43 + 5 [ ] 50   c. 3 + 33 [ ] 36

THE PLANT MAZE

Help the Princess find her way back to the Castle.

84

What is the number of the day?

Write the number using words:

Write the number three different ways using sticks and circles.
(Remember a stick represents a ten and a circle represents a one. If you know how to use a square block to represent a hundred, you may do so.)

Adding and Subtracting Two-Digit Numbers

8. Add.

a. 84 + 4 = _____	b. 6 + 70 = _____	c. 74 + 5 = _____
41 + 4 = _____	16 + 2 = _____	6 + 53 = _____

9. Subtract.

a. 80 – 30 = _____	b. 55 – 3 = _____	c. 29 – 3 = _____
17 – 3 = _____	100 – 40 = _____	50 – 2 = _____

Can you find
the way to the
picnic basket?

Will you help Timmy find his twin brother Tommy?

What is the number of the day?

Write the number using words:

Write the number three different ways using sticks and circles.
(Remember a stick represents a ten and a circle represents a one. If you know how to use a square block to represent a hundred, you may do so.)

10. Add and subtract.

a.   1 4
    + 3 5
    ‾‾‾‾‾‾

b.   5 9
    − 3 4
    ‾‾‾‾‾‾

c.   4 0
    + 5 6
    ‾‾‾‾‾‾

d.   9 6
    − 6 0
    ‾‾‾‾‾‾

11. Add. The images can help you.

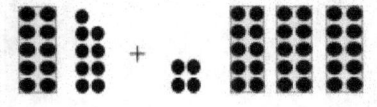 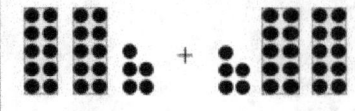

a. 19 + 34 = _____

b. 25 + 25 = _____

c. 22 + 27 = _____

Basic Word Problems

12. Write a subtraction sentence that matches with the addition 6 + 8 = 14.

_____ − _____ = _____

13. How many more is 70 than 50?   _____ more

14. Henry owns four more cars than Mark, and Mark owns six cars.
    Draw Mark's cars and Henry's cars.

15. Ten kids are playing in the yard. There are 6 boys. How many girls are there?

Skip Counting by 2

Directions: Help the penguin chick find her mom. Skip count by two and complete the maze. Color in the squares as you count.

1	55	7	43	75	6	11	25	44	75
65	4	66	4	54	65	46	48	50	52
20	22	24	26	28	64	44	85	8	54
18	65	90	85	30	59	42	19	2	56
16	14	12	64	32	74	40	41	3	58
61	54	10	78	34	36	38	28	64	60
4	6	8	75	52	35	5	84	79	25
2	5	25	27	94	64	28	16	38	57
5	58	85	14	75	35	29	11	52	87
56	25	75	95	13	35	94	46	66	25

2-60

Skip Counting by 2

Directions: Help the penguin chick find her mom. Skip count by two and complete the maze. Color in the squares as you count.

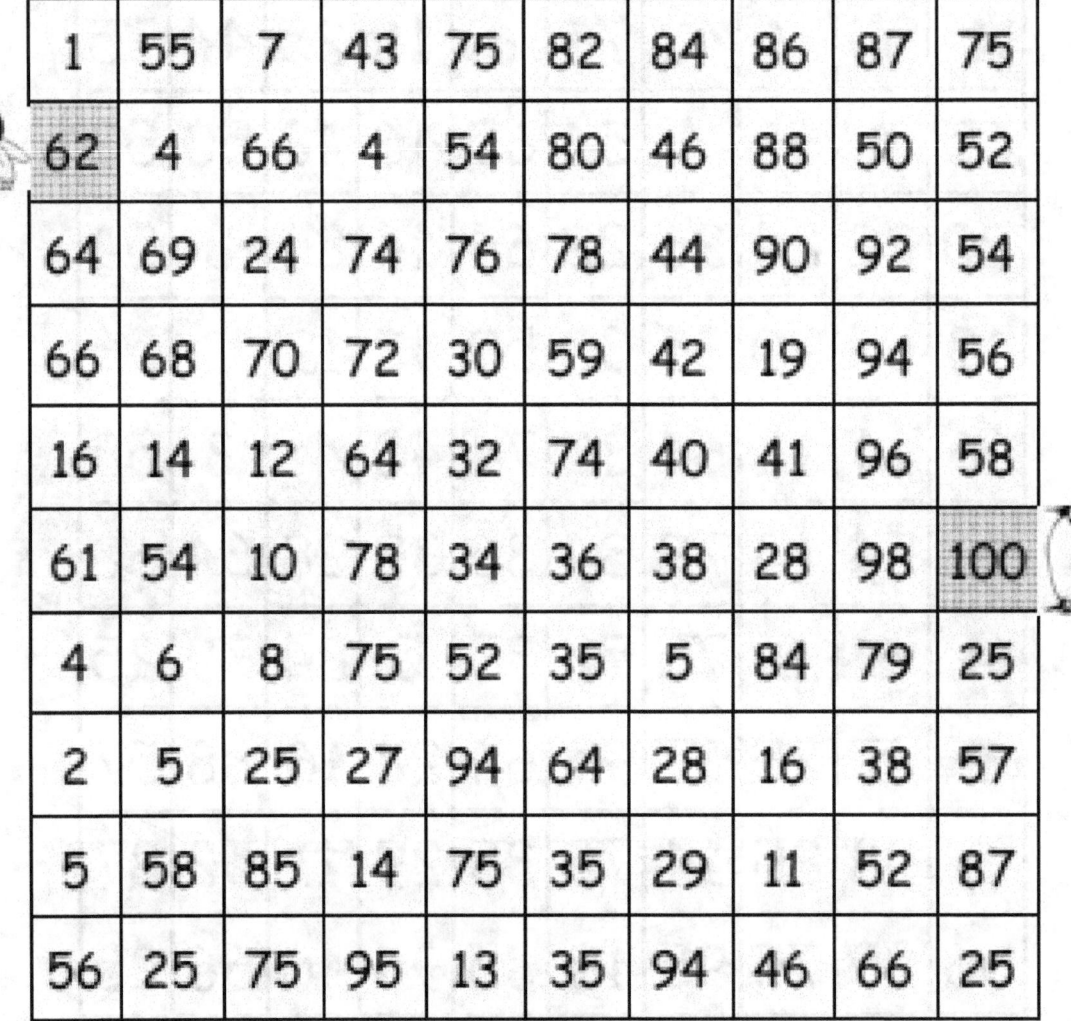

1	55	7	43	75	82	84	86	87	75
62	4	66	4	54	80	46	88	50	52
64	69	24	74	76	78	44	90	92	54
66	68	70	72	30	59	42	19	94	56
16	14	12	64	32	74	40	41	96	58
61	54	10	78	34	36	38	28	98	100
4	6	8	75	52	35	5	84	79	25
2	5	25	27	94	64	28	16	38	57
5	58	85	14	75	35	29	11	52	87
56	25	75	95	13	35	94	46	66	25

62-100

Skip Counting by 5

Directions: Help the penguin chick find her mom. Skip count by two and complete the maze. Color in the squares as you count.

1	55	7	43	75	82	84	86	5	75
62	4	66	4	54	80	46	88	10	52
64	69	24	74	76	78	25	20	15	54
66	68	70	72	30	59	30	19	94	56
16	14	12	64	32	74	35	41	96	58
100	95	90	85	34	36	40	28	98	100
4	6	8	80	52	35	45	84	79	25
2	5	25	75	94	64	50	16	38	57
5	58	85	70	65	60	55	11	52	87
56	25	75	95	13	35	94	46	66	25

5-100

Skip Counting by 5

Directions: Help the hummingbird find a flower do drink from. Skip count by five and complete the maze. Color in the squares as you count.

1	200	7	43	75	82	84	86	5	75
62	195	190	185	180	80	46	88	10	52
64	69	24	74	175	78	25	20	15	54
66	68	70	165	170	59	30	19	94	56
16	14	12	160	32	74	35	41	96	58
100	95	150	155	34	36	40	28	98	100
4	6	145	80	52	35	45	84	79	25
2	5	140	135	130	125	120	115	110	57
5	58	85	70	65	60	55	11	105	87
56	25	75	95	13	35	94	46	100	25

100-200

Skip Counting by 10

Directions: Help the leprechaun find his pot of gold. Skip count by 10 and complete the maze. Color in the squares as you count.

1	55	7	300	75	82	84	86	5	75
62	4	66	290	280	270	46	88	10	52
64	69	24	74	76	260	25	20	15	54
66	68	70	72	30	250	240	230	220	210
16	14	12	100	110	120	130	41	96	200
100	95	90	90	34	36	140	28	98	190
4	6	8	80	70	35	150	160	170	180
2	5	25	50	60	64	50	16	38	57
5	20	30	40	65	60	55	11	52	87
56	10	75	95	13	35	94	46	66	25

10-300

95

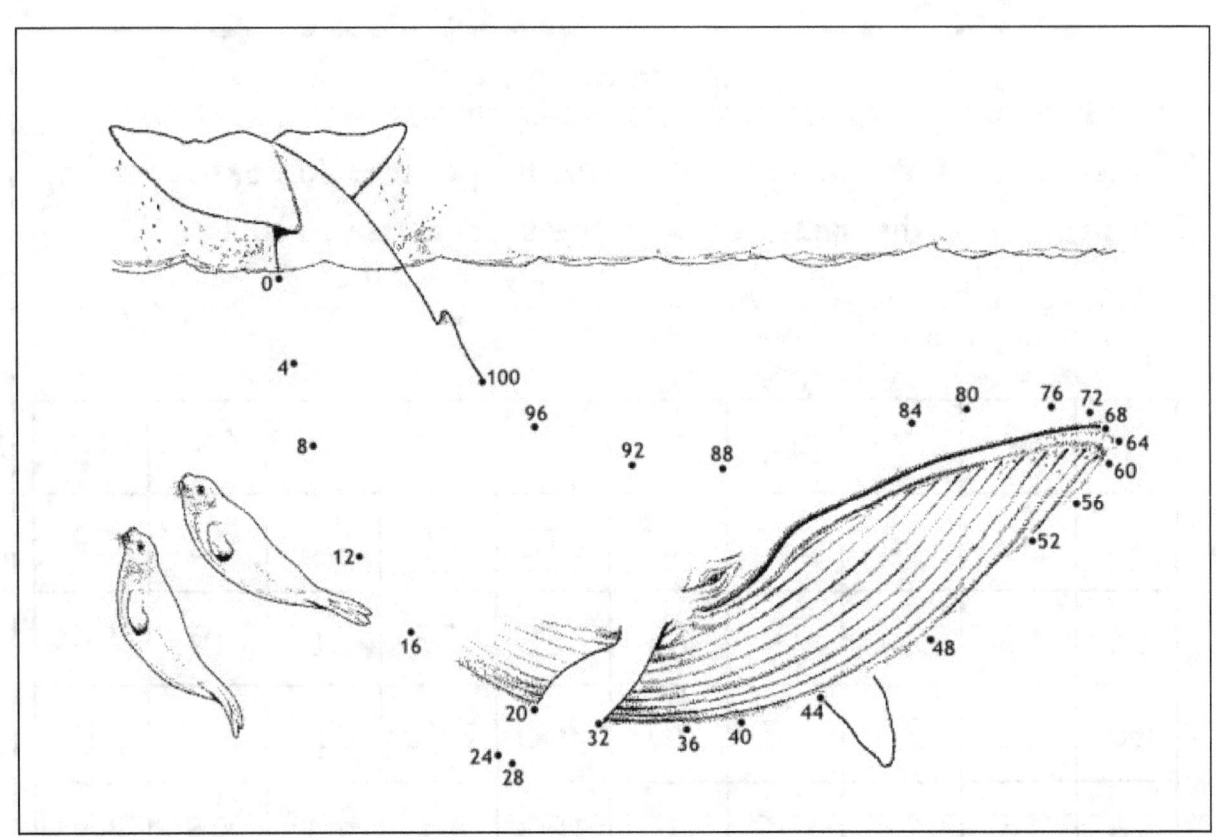

Color to make a pattern.

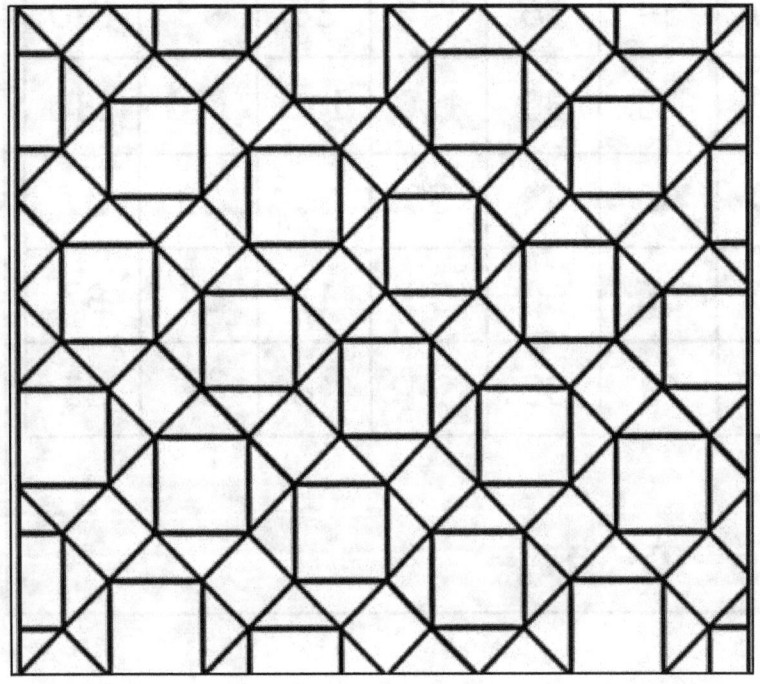

Color to design a pattern.

Color to design a pattern.

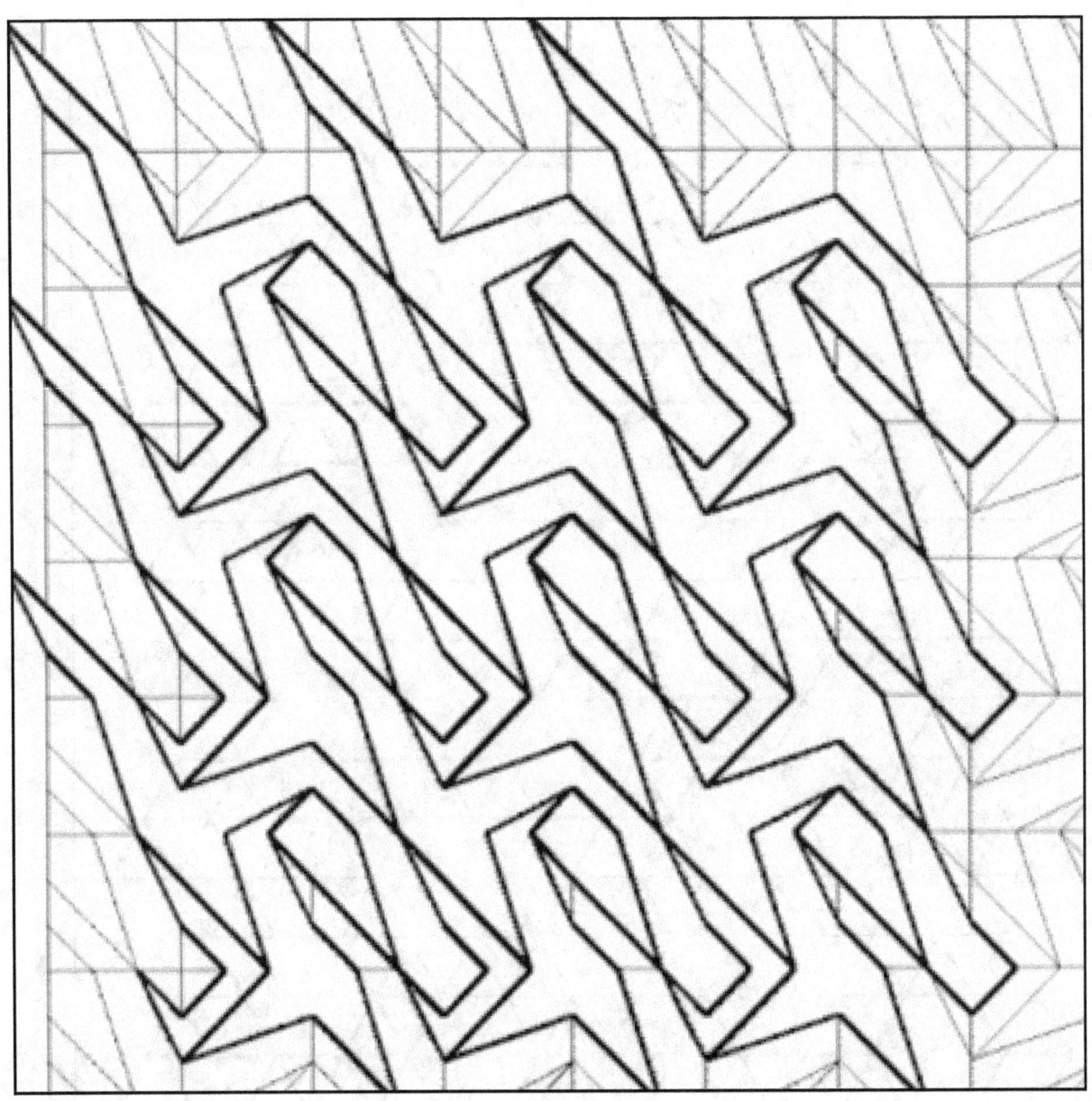

Color to design a pattern.

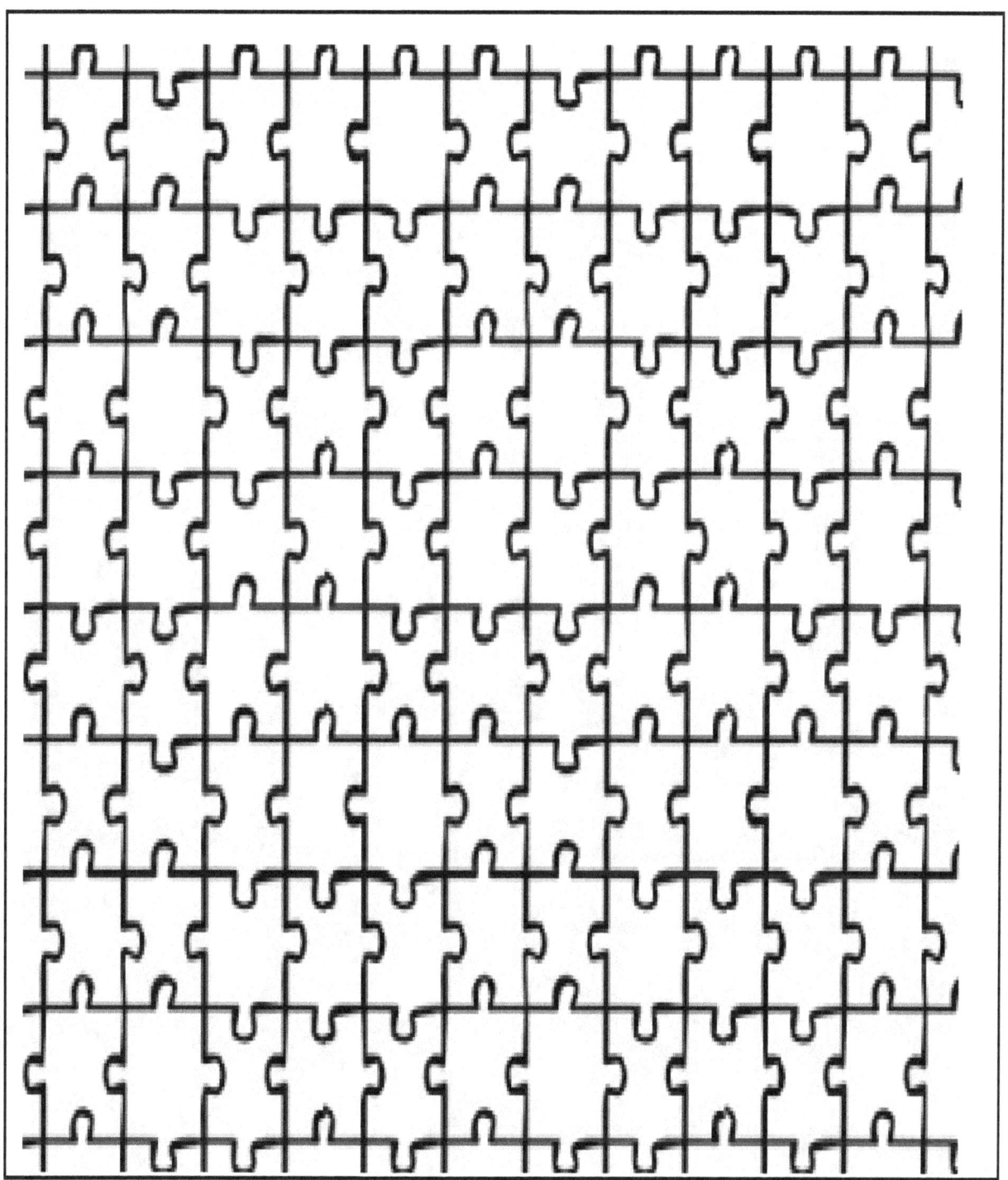

Free Drawing Pages

www.ingramcontent.com/pod-product-compliance
Lightning Source LLC
Chambersburg PA
CBHW081737220526
45468CB00008B/2133